Beekeeping and conserving biodiversity of honeybees

Sustainable bee breeding
Theoretical and practical guide

Northern Bee Books

Beekeeping and conserving biodiversity of honeybees
Sustainable bee breeding
Theoretical and practical guide

Editors: Marco Lodesani & Cecilia Costa

Publication was partially supported by the Commission of the European Communities. EC program EESD - Project Contract n. EVK2-CT-2000-00068, Beekeeping and Apis Biodiversity in Europe (**BABE**)

Originally published in July 2005 by Northern Bee Books

Reprinted in 2011

and available from:

Northern Bee Books

Scout Bottom Farm

Mytholmroyd

Hebden Bridge HX7 5JS (UK)

www.groovycart.co.uk/beebooks

ISBN: 978-1-904846-14-7

Beekeeping and conserving biodiversity of honeybees

Sustainable bee breeding
Theoretical and practical guide

Contents

CHAPTER 2

Selection theory and effective population size.....................53

Michel Solignac, Jean-Marie Cornuet

PART II   SPECIAL PROJECTS

CHAPTER 5

Honeybee conservation: a case story from Læsø Island (DK)..142

Annette Bruun Jensen, Bo Vest Pedersen

Contributors (in alphabetic order)

- Jean-Marie Cornuet

Centre de Biologie et de Gestion des Populations Campus International de Baillarguet

CS 30016 Montferrier sur Lez 34988 Saint Gely du Fesc Cedex, France

E-mail: jmcornuet@ensam.inra.fr

- Cecilia Costa

Istituto Nazionale di Apicoltura, via di Saliceto 80, 40128 Bologna, Italy.

E-mail: c.costa@stpa.unibo.it

- Pilar De la Rúa

Departamento de Biotecnología Instituto Murciano de Investigación y Desarrollo Agrario y Alimentario (IMIDA) C/ Mayor s/n, 30150 La Alberca, Murcia, Spain.

E-mail: pdelarua@correo.um.es

- Stefan Fuchs

Institut für Bienenkunde (Polytechnische Gesellschaft) Fachbereich Biologie und Informatik der J. W. Goethe-Universität, Frankfurt am Main Karl-von-Frisch-Weg 2, 61440 Oberursel, Germany

E-mail: s.fuchs@em.uni-frankfurt.de

- Annette Bruun Jensen

Institute of Biology, University of Copenhagen, Universitetsparken 15, DK 2100 Copenhagen Ø.

E -mail: anbjensen@bi.ku.dk

- Marco Lodesani

Istituto Nazionale di Apicoltura, via di Saliceto 80, 40128 Bologna, Italy.

E-mail: m.lodesani@stpa.unibo.it

- Robin Moritz

Institut für Zoologie, Martin-Luther-Universität Halle-Wittenberg

Hoher Weg 4, D 06099 Halle/Saale, Germany

E-mail: r.moritz@zoologie.uni-halle.de

- Bo Vest Pedersen

Institute of Biology, University of Copenhagen, Universitetsparken 15, DK

2100 Copenhagen Ø.

E -mail: bvpedersen@bi.ku.dk

- José Serrano

Departamento de Zoología Facultad de Veterinaria, Universidad de Murcia

Campus de Espinardo.   Aptdo. 4021, 30071 Murcia (Spain).

Email: jserrano@um.es

- Michel Solignac

Laboratoire Populations, Genetique et Evolution Université Paris Sud

(Orsay) Avenue de la   Terrasse 91190 Gif-sur-Yvette, France

E-mail: Michel.Solignac@ pge.cnrs-gif.fr

BEEKEEPING FOR MAINTAINING BIODIVERSITY

Robin F.A. Moritz

1. THE HONEYBEE, A DOMESTICATED ANIMAL?

The honeybee, *Apis mellifera**, is the world's most important beneficial insect. Honeybees are of great socio-economic value to man. They produce honey, pollinate crops worth billions of Euros per year, and provide full time and part time employment to beekeepers throughout the globe. The honeybee is also a vital member of many terrestrial ecosystems via its pollination activities. As a generalist it pollinates a broad spectrum of wild flora also early in spring when other solitary pollinators are not yet present. Because they hibernate as colonies with large numbers of workers they are essential for ensuring early spring pollination. Although bumblebees have been show to be more effective in pollination as individuals, this does not compensate for the large numbers of honeybees at the beginning of the season.

Honeybees are often considered as "domesticated" animals. However this is not true. Man managed to keep honeybees in boxes but is far away from domesticating this insect as poultry, pork or cattle are. Honeybees can leave their hive at any time to successfully reproduce in the wild. In spite of all breeding efforts and development in apiculture, beekeeping is more like aquaculture, where fish are bred in ponds or estuaries. No trout fisher would consider trout a domesticated fish just because it is bred in ponds. The honeybee is a wild animal that we keep and utilise while it resides in the artificial nest site we provide allowing access to its resources.

Equally it is often confusing when people refer to colonies not kept in hives as "feral" colonies. In regions where the honeybee occurs naturally, it is much more appropriate to refer to these as "wild" colonies. Honeybees in Europe, Africa, and the Middle East of Asia are vital parts of ecosystems and essential for their functioning. The discrimination between "wild" bees as all non-honeybee Apoidea* is particularly misleading and fundamentally wrong. Most honeybees are wild bees and they are among the most essential pollinators we know. Conservationists often regard the honeybee as vermin or pest eradicating precious rare species or spreading weeds. We will see below that these opinions primarily result from "gut feeling" rather than empirical evidence. This book is a pledge to those involved in conservation of natural habitats to respect the evidence and consider the honeybee as a native wild bee and a not priory as an additional biodiversity threat from agricultural activities. It may be true that massive assemblies of honeybee colonies of migratory beekeepers can cause competition with other pollinators. However, the evidence for actual competition between honeybees with other bees is usually lacking. This is because experimental approaches are extremely difficult [1]. Even in the most dramatic experiment: the invasion of the African honeybee in the Americas, which has been scrutinized for many decades, there are no reports of extinctions of pollinators. For honeybees in the Americas it indeed is appropriate to use the term "feral" populations, because man brought the honeybee to the American continent (as well as to Asia and Australia). Clearly many extremely plausible arguments can be made as to why and how honeybees should be more efficient in foraging and as a consequence in heavy competition to other pollinators. Honeybees communicate and can recruit to nectar sources. They forage over much larger distances than other bees. They start earlier in the day with foraging, thereby depleting floral resources for others. Competition can certainly be possible and several cases have been reported where indeed other bees shifted in the floral forage spectra when honeybees where introduced [2]. However, does it matter at the ecosystem level? Here the evidence is lacking, and until to date there are no rigorous reports of any extinctions as a response to honeybee introductions [1,3]. Indeed the potential for conflict with other pollinators may be less dramatic as it may look at first glance. Honeybees have completely different nesting requirements than other bees. They do not forage aggressively as do several stingless bees* species [4] and can rapidly shift the foraging force to non-competitive nectar sources.

There are reports for the opposite. The removal of feral honeybee* colonies from Santa Cruz Island in California resulted in an increase of other flower visiting insects [5], but this again shows that honeybees did not cause extinctions of the native pollination fauna on this island.

2. CONSERVATION OF WILD HONEYBEES

Most people will agree that honeybees are of great importance to man, but only few realise that it also requires conservation in those regions where wild populations still exist. How is it that a single generalist* species that is widespread (native to Africa, Europe and the Middle East (see Chapter 1), and introduced to Asia, America and Australia requires conservation? The reason is that natural honeybee populations in Europe and Africa (less so) have been seriously affected by human activities. Man has transported and propagated non-native races of honeybees around the globe. Wild honeybee populations also compete with managed hives for flowers, face numerous new diseases introduced by man. The most dangerous but often disregarded risk for biodiversity stems from excessive gene flow from vast numbers of managed colonies into a decreasing wild population. So the biodiversity risk lies less in interspecific competition* rather then within species gene flow between managed and wild populations. Due to the specialised mating biology, the honeybee is the only agricultural animal where the man kept population shares its gene pool with the wild population. Queens mate in mid air at drone congregation areas with drones from colonies as far away as 10 km causing gene flow between wild and commercially kept populations.

Bee breeding in the 20th century was dominated by introducing "superior" honeybees from various parts of Europe and Africa into commercial beekeeping. This practice overlooks the importance of local adaptation and disregards the need for conserving biodiversity. As a consequence, native honeybees are considered to be extinct in many parts of Europe. From a conservation perspective two important components of honeybee diversity are threatened:

1) Native races and subspecies of honeybees, which are adapted to their local environment;

2) Genetic diversity within local populations.

This genetic diversity is of great importance both as part of Europe's natural biological heritage and as a source of genetic variation for the continued use of the honeybee in agriculture. For example, among the genetic variation lost may be resistance to diseases (such as Varroatosis and American Foulbrood) and other desirable traits.

This book will present the biological background, the beekeeping technology and the bee breeding tools for a sustainable bee management with locally adapted endemic* honeybees. It will show both the professional and the amateur beekeeper how to establish breeding strains within an environment potentially contaminated with imported honeybee stock. The goal is to reconcile the interest of beekeepers with the need for conserving endemic honeybee populations for beekeeping.

COMMERCIAL APICULTURE VS. HOBBY BEEKEEPING VS. CONSERVATION?

Clearly the apicultural industry needs honeybees that match their needs. These are manifold and depend on the tasks the bees are used for. Honey producers are clearly in need of productive and healthy colonies and traits like defensive behaviour are less important. Those beekeepers using honeybees for pollination will require colonies that rapidly built up in spring. Amateurs will prefer docile honeybees which allow the to work on even in densely populated urban environments. Queen breeders therefore will have great interest in selling less defensive stock if their main market lies in supplying amateur apiarists. All have one goal in common. They will prefer to work with bees that are healthy and do not require specific treatment for diseases.

In light of the different purposes it is not surprising that different subspecies and breeds are preferred by different beekeepers. So, how can we possibly reconcile the diverse interest of the industry, the hobbyist and conservation? This is not easy indeed and this is what this book is all about. We will explain out how professional beekeepers can maintain their preferred breeding lines yet respecting the needs of conserving endemic races. We will explain what hobby beekeepers can do to promote and actively engage in the conservation of their endemic bees. We will show how commercial beekeepers can profit on the long run by using endemic honeybees rather than relying on pre-selected stock obtained somewhere on the globe.

BREEDING FOR CONSERVATION

Many tools already are routine procedure in apiculture although they have been developed for very different reasons. Separating wild populations from selected breeds has been apicultural practice in many European countries. This was usually achieved through mating control on islands or remote honeybee free regions were virgin queens and drones were placed for mating. Quality control of matings was based on morphometric measurements, which each beekeeper could easily perform himself. For example generations of German beekeepers screened for the cubital index to discriminate the desired carniolan race *Apis mellifera carnica** from the endemic "black" bee *Apis mellifera mellifera**. The cubital index (fig. 1) and a few other morphological characters provided a safe diagnostic tool to separate selected stock from wild honeybees and hybrids.

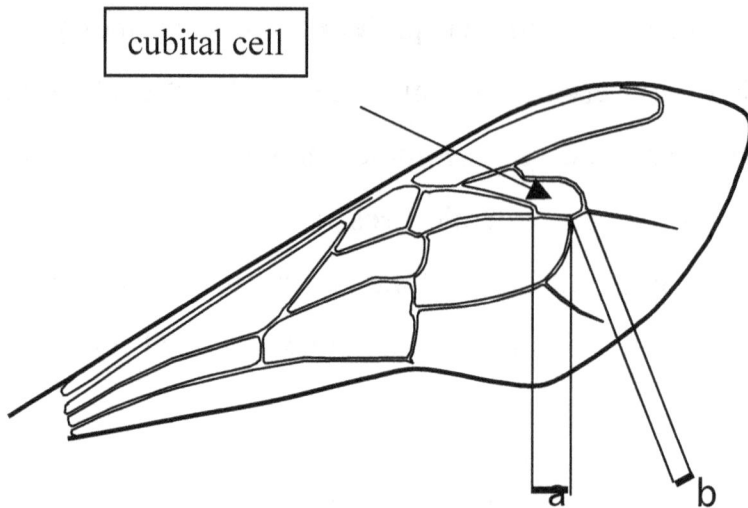

cubital cell

Figure 1 : **The cubital index* (a : b) in the wing venation pattern is the main character discriminating** *A. m. mellifera* **from** *A. m. carnica.* **It is used as a tool by beekeepers to identify mismatings of carniolan* queens with** *mellifera* **drones.**

The cubital index* in *A. m. mellifera* is much smaller than in *A. m. carnica.* If the cubital index of the worker offspring of a given queen was too low, this queen was discarded from the *carnica-*breeding program irrespective of other traits. More recently molecular DNA techniques proved useful to discriminate between races (see chapter 1), however these are of more use for scientists rather than the beekeeper.

The regulatory framework that has been developed for running honeybee-mating stations for breeding honeybees for commercial purposes cannot directly be transferred to running conservation schemes. Mating stations have very different requirements than those currently used for commercial honeybee breeding (see chapter 3). They need to be larger, need good climatic conditions and must accommodate a large number of colonies throughout the season.

Beekeepers engaged in conservation efforts are by no means freed from careful selection work. The opposite is true apart from selecting for desired traits he has also take into account the genetic diversity in his population. This causes additional work and costs. The additional goal in breeding for conservation lies in maintaining diversity rather than selecting for a specific colony phenotype that promises best economic returns. Breeding must also aim to maximize effective population sizes (chapter 2) and maintain genetic variability. Classical breeding and population genetic theory needs to be drastically modified to take into account that honeybees live in colonies. Matters are complicated because queens are diploid*, mate with large numbers of drones, but drones are haploid* (carry only half the genetic information as females do). Chapter 2 highlights in detail how to genetically deal with a colony trait that is composed of about forty thousand individual genomes, which constantly interact with each other, which are from different generations, which belong to different castes. The caste* system itself (with sterile workers and a fertile queen) poses an additional complication to unravelling the genetics of a honeybee colony.

Queen and drone rearing are routine procedures in apiculture. Yet again the procedures need to be adapted if breeding programmes are run for conservation as a primary breeding goal. Mating must be as diverse as possible to take advantage of the mating system of the honeybee. Chapter 4 deals in detail with the peculiarities of running a breeding facility for conserving endemic populations.

ORGANIC BEEKEEPING

Finally we show in two case studies how conservation of honeybees is not only an important service to global conservation of biodiversity, but also a story of sustainable economic success. We predict that this economic success will increase as consumers of bee product value local produce of specific regions. Regional products can become a brand where customers not only appreciate the product from a specific region but also the animals producing the honey. Beekeepers can reach both better prices and wider markets just by using endemic honeybees rather than imported stock. EU regulations are already in place to support this development by having a directive for "organic beekeeping", which is based on using indigenous honeybees rather than imported stock. It is worth while studying the Council regulation (EC) No 1804/1999 of 19 July 1999 supplementing Regulation (EEC) No 2092/91 on organic production of agricultural products explicitly addresses honeybees. It shows that the value of regional products and organic production has been acknowledged at the political level and is actively supported. In paragraph nine of this regulation we read on p.1

(9) *A wide biological diversity should be encouraged and the **choice** of breeds should take account of their capacity to adopt to **local conditions**.*

and further down in Annex 1 on p 9

3.1 *In the choice of breeds or strains, account must be taken of the capacity of animals to adapt to local conditions; their vitality, and their resistance to disease. In addition, breeds or strains of animal shall be selected to avoid specific diseases or health problem associated with some breeds or strains used in intensive production Preference is to be given to **indigenous breeds and strains**.*

It is therefore very clear that organic farming including apiculture must rely on local breeds and not stock imported from all around the globe. Selective work is encouraged breeding can explicitly include artificial insemination as we can further read on p 13

6.1.1. *In principle, the reproduction of organically reared livestock should be based on natural methods. Nevertheless **artificial insemination is permitted**. Other forms of artificial or assisted reproduction (for example embryo transfers) are prohibited.*

In section C starting on p17, which deals exclusively with apiculture, we find

Origin of the bees

3.1. In the choice of breeds, account must be taken of the capacity of animals to adapt to local conditions, their vitality and their resistance to disease. Preference shall be given to the use of European breeds of Apis mellifera and their local ecotypes.

A few other peculiarities of the organic beekeeping regulation may be listed, to illustrate that it goes far beyond the goals for conserving indigenous populations proposed in this book.

7.2. Mutilation such as clipping the wings of queen bees is prohibited.

This probably well meant regulation has clearly no bearing on neither well-being of the queen, honey quality, nor on conserving biodiversity.

Nevertheless, given the value European societies gives to organic apiculture and the sometimes surprising detail policy makers devote to making regulations, it is obvious that measures must be put into place to allow for sustainable functioning of such operations. At the moment efforts are incipient at best in most European countries. In vast regions organic apiculture is simply impossible because needed restrictions on bee movements are not in place allowing for uncontrolled gene flow between potential organic beekeepers and others that prefer to work in classical apiculture.

Acknowledgement

This book was made possible through an EU research network on the impact of beekeeping on honeybee diversity (BABE network). We particularly appreciate the help and advice of Martin Sharman (DG 12) in making BABE a successful network. His interest went far beyond his regular administrative duties and throughout the duration of the project he consistently assisted in making the network a success.

References

[1]   Goulson D. (2003) Effects of introduced bees on native ecosystems. Ann. Rev. Ecol. Evol. Syst. 34,1-26.

[2]   Roubik D.W. (1978) Competitive interactions between neotropical pollinators and Africanized honey bees. Science 201, 1030-1032.

[3]   Roubik D.W., Wolda H. (2001) Do competing honey bees matter? Dynamics and abundance of native bees before and after honey bee introduction. Popul. Ecol. 43, 53-62.

[4]   Roubik D.W. (1991) Aspects of Africanized honey bee ecology in tropical America. in: Spivak M., Fletcher D.J.C.,

Breed M.D. (eds.) The "African" Honey Bee. Westview Press, Boulder CO. pp 259 - 281.

[5]   Wenner A.M., Thorpe R.W. (1994) Removal of feral honey bee (*Apis mellifera*) colonies from Santa Cruz Island. in: Halvorson W.L., Maender G.J. (eds) The Fourth Californian Islands Symposium: Update on the Status of Resources. Santa Barbara Museum of Natural History, Santa Barbara CA, pp 513-522.

[6]   Hemila K (1999) Council Regulation EC No 1804/1999 Off. J. Eur. Comm. L222, 1-28.

CHAPTER 1

BIOGEOGRAPHY OF EUROPEAN HONEY BEES

Pilar De la Rúa, Stefan Fuchs, José Serrano

1. MORPHOLOGICAL RACES* OF THE HONEYBEE *APIS MELLIFERA*

The natural habitat of *A. mellifera* covers an enormous distance of more than 10000 km from South Africa to the northern parts of Europe. Considering the variety of climatic zones, honey bees are nevertheless fairly uniform in appearance. This is likely due to some degree of independence from outside climatic conditions attained through the homeostatic properties of colonial life, in particular the pronounced ability to regulate temperature. Nevertheless it has been recognised for long times, that bees do not only show considerable differences in behaviour and colonial conduct, but also in morphological traits. Of these, variation in colour is most conspicuous, but also size, and, on closer inspection, size relations of different body parts and patterns of wing venation show persistent differences which are characteristic for particular regions or bee strains. In the general trend, bees tend to be bigger and stouter in the cool climates at increasing geographic latitude or elevation above sea level, which is in accordance to the general rules of Bergmann* and Allen* pertaining to relation of geography and animal morphology.

Bees also tend to be darker in the cooler climates, which runs contrary to Gloger's rule*. However, many exceptions stress the importance of particular local conditions and the history of spread.

For the purpose of systematic bee breeding convenient methods for distinguishing and characterising bees are required. In comparison to behaviour or traits of colonial development, bee morphology is easy to assess, and can be broken down into the exact measurements of particular morphometric traits, e.g. the length of wings or proboscis*, patterns of wing venation, or pigmentation.

Morphometry* has a long history in honey bee research. Considering the only slight and continuous variation, characterisation needs to be based on quantitative measurements rather than the description of qualitative traits. As early as 1928 the Russian investigator [1] introduced measurements of various traits as length of proboscis*, etc. into bee breeding experiments. During the systematic replacement of the local bees of Germany during the last century, distinguishing bees by measurements of wing venation patterns, most notably by the cubital index*, came in broad use in practical bee breeding [21].

Ruttner et al. [24] integrated several established and new measurements into a system of 36 morphometric traits (15 sizes, 5 colours, 13 wing venation and 3 hair characteristics). At the same time, the work scope expanded into an inventory of honey bees world wide. This was prompted by analysing the collection of

Brother Adam which covered the main regions of beekeeping, and was later complemented with additional samples coming in over time. In 1988 Ruttner [23] presented an overview on biogeography of *A. mellifera*, based on the methods of quantitative taxonomy*, which still stands out as a standard reference for subspecies variation in honey bees as well as in terms of the measuring system itself (Rutttner, 1988). At present, the Oberursel data bank contains about 1600 samples from all over Europe.

To visualise the morphological variation in geographical space, samples are mapped in figure 1. The spatial distribution of sample colours shows a fairly clear substructure of the honey bee population in Europe and adjacent regions. High admixtures of red indicate large bees, occurring north of the Alps, France and Spain; admixtures of blue indicate yellow-coloured bees, as in the Near East or Italy.

The geographic pattern emphasises the role of natural history in the morphology of the bees. Ruttner et al. [24] hypothesised that honey bees invaded the European and African continent presumably from the near-East after "inventing" advanced thermoregulation, splitting into four major branches, one invading Africa, one invading the Balkans and Italy, one taking a more eastern route propagating north of the Alps to France and parts of Spain, while the fourth branch ended in near east (Anatolia and Iran).

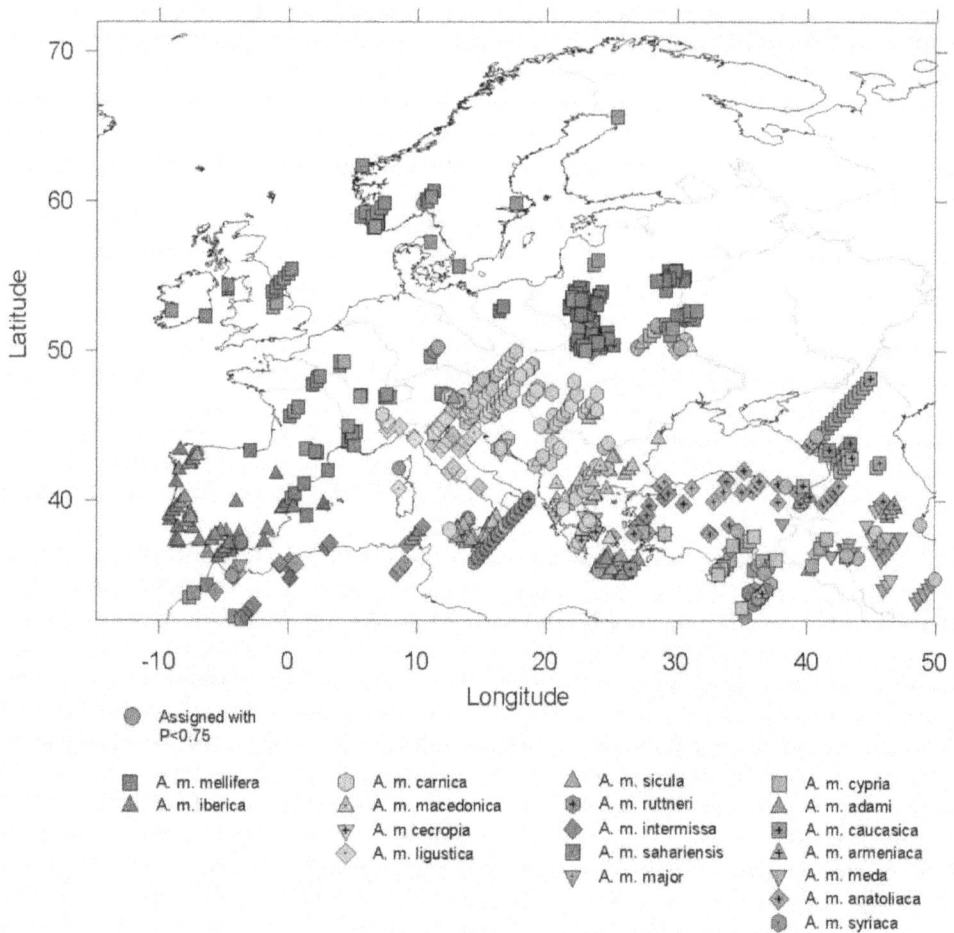

Figure 1 : **Distribution of 830 samples of *A. mellifera*. Each point represents one colony sample; samples from the same location are stacked. Colours of samples were created from factor analysis scores as in Fig. 1 as an aid to visualise morphological trends. Symbols represent different subspecies determined by discriminant analysis, the respective subspecies symbols are given below the graph ordered by the main branches M, C, A and O, with colours derived from mean factor values for the branches.**

(This illustration in the original publication was in colour which was not available in this edition)

18

The current distribution has been much influenced by re-colonisation of the continent from Mediterranean refuges some 8000 years ago, to which bees have been restricted during the last ice ages. However, these colonisation processes did not change significantly the distribution of these main branches, first discovered by morphometry* and later confirmed by DNA methods (see below), as they can be clearly verified from sample colour distribution in the map. In particular the distinct boundaries between Anatolia and Greece, both with similar climates, reflect the historical barriers of spread of different branches rather than local adaptations to climate. Islands may provide stepping stones, and colour mixtures clearly indicate both Balkanian and Anatolian influences on the Greek Islands and clear influences of African characteristics in Malta and Sicily.

1.1. Subspecies or races*

The species *Apis mellifera* Linné, 1758 includes many subspecies or geographic races that are named using a subspecific name, e.g., *Apis mellifera ligustica*. Each race represents a certain degree of differentiation in all types of characters (genes, morphological, ecological, etc.) that has been possible because of a certain degree of geographic and genetic isolation of the race. While local populations may show constancy in some racial traits that permit their classification into a given subspecies, clear-cut boundaries between the races are often difficult to trace on a larger geographical scale. This difficulty is even more apparent if characters show a graduated spatial distribution (i.e., a cline) as opposed to definite boundaries. In addition to these objective problems, the subdivision of a species into subspecies is also influenced by subjective elements, as are the perception of the investigator about the degree of difference, or when and where a variation was first recognised. Thus, to name variation in *Apis mellifera* below the species level is widely a matter of convenience for practical purposes and lacks adequate biological definition.

Over time, numerous subspecies names have been in use. In his monograph on *The Biogeography of Honey Bees*, Ruttner [23] took a conservative approach by adopting 25 of them (Ruttner admitted 24 and was doubtful about *A. mellifera major*), of which 9 are European. Of each subspecies, Ruttner selected about 10 to 20 reference samples, most stemming from Brother Adam's collections. These then defined the subspecies' morphological properties. Further samples in the Oberursel data bank were then allocated to these subspecies by discriminant analysis, a statistical procedure that finds the most useful measurements for differentiating the subspecies, and determines the likelihood samples belonging to these. In a detailed taxonomic revision Engel [11] admitted 28 subspecies, including some newly described ones, and corrected the names and authorship of some of them.

The geographical pattern of subspecies allocation is also shown in figure 1. Mostly, samples allocated to one subspecies cover coherent regions, coinciding with those they have been described for. However, regions particularly between the main areas of distribution may show some overlap and uncertain allocation (confidence < 75%, circles), pointing to graded character distributions opposed to sharp boundaries. Stray samples are likely to occur as a consequence of bee imports, which have disturbed the natural patterns of subspecies distribution.

The resulting geographic pattern of subspecies again reflects the concept of four main branches and also coincides with the distribution of the colour coding. However, within the major branches transitions are often subtle. Subspecies definitions thus show some degree of arbitrariness, imposed on gradual character shifts due to migration or graded selection rather than anything which easily lends itself to be boxed and named.

Subspecies, even if unambiguously characterised, still may encompass a wide degree of variation. Some span extended regions with distinct climatic differences, which again have led to marked differences, sometimes as pronounced as differences between subspecies. This puts limits to generalisations on traits of bees of particular subspecies, and in view of such differences the specific locality a bee used for breeding or research, is as important for its characterisation as any subspecies allocation.

Notwithstanding the possible shortcomings, the subspecies system proposed by Ruttner [23] has become a concise reference as it is based on comparative measurements and provides information not only on morphological traits but also on behavioural inclinations of European honey bees. It is a frame of invaluable help for further improvement of understanding.

1.2. A short subspecies overview

1.2.1. M Branch

The M branch covers most of Europe, from Spain to the Ural, except of the regions southern to the Alps, and from there to the east southwards of about 50° latitude. The bees are comparatively large and dark coloured, with broad abdomen. They are sparsely haired, with a tendency for defensiveness and use of propolis.

The two subspecies, *A. m. mellifera and A. m. iberiensis* are separated by the Pyrenees. By its characteristic ability to cope with cold and unfavourable climates, *A. m. mellifera* has colonised all temperate and northern regions of Europe, probably spreading out from Iberia and SW France refuges after the last ice age. Generally, the seasonal brood production and colony development start out slowly only late in spring, with the main emphasis on collecting late flowering plants, of which heather is prominent in many regions. Though a clearly defined subspecies, it shows a fair amount of variation due to local ecotypes* (Black Bee "nigra" in the Swiss Alps, "Brown Bee" in Norway, showing superior thermoregulation). In particular in the eastern regions variation is not well documented but is indicated by proposed subspecies names as *A. m. silvarum* for the bees of eastern Poland and the central regions of Russia. Its role in beekeeping is declining in many regions. Particularly in Germany, the bee has been almost completely replaced by other more gentle and fast breeding bee subspecies.

A. m. iberiensis in Portugal and Spain is generally smaller and shorter haired than *A. m. mellifera,* and the morphometric distinction between both subspecies is gradual in the northern half of the Iberian Peninsula. Ruttner [23] noted that this subspecies shows characteristics of *A. mellifera intermissa*, the bee of North Morocco included in the African branch. Additionally to morphological characteristics there are behavioural ones, a quick defence reaction, nervousness on the comb, propensity to swarm, and ample use of propolis [23]. On the other hand, *A. m. iberiensis* shows a similar cubital index* (CI) and the ability to survive to cold and long winters that relate this subspecies to *A. mellifera mellifera.*

1.2.2. C Branch

The C branch encompasses six subspecies. Four of them are distributed in the Apennine Peninsula, the Balkans and South Ukraine: *A. m. ligustica, A. m. carnica, A. m. macedonica, A. m.*

cecropia. The other two are found in Sicily and Crete (see below). The C branch bees are smaller than the M branch bees, of variable colour from grey to distinctly yellow, with shorter hairs and a high cubital index*. No distinct boundaries exist between the bees of the Balkans, *A. m. carnica, A. m. macedonica* and *A. m. cecropia,* and the ecological richness of the region corresponds with the pronounced variation within these subspecies. While capturing the main trends over the region, these 3 subspecies may not represent separate clear-cut groups. *A. m. carnica* contains at least two distinguishable ecotypes* (pannonic and alpine), and morphometric investigations in the Balkans continue to identify separable subpopulations.

A. m. ligustica, the Italian bee, is clearly set apart mainly by its distinctly yellow coloration and amalgamation with *A. m. carnica* is limited to a small zone in the North of Italy. Though belonging to the C branch, *A. m. sicula,* the bee of Sicily, shows strong African influence in morphometry* and behaviour, as happens with *A. m. iberiensis.*

A. m. carnica and *A. m. ligustica* are the most successful subspecies used in beekeeping. Their rapid colony build up in spring and successful breeding for gentleness makes them useful for exploiting early spring honeyflows and easy to handle. *A. m. carnica* in its mountain varieties proved adaptable to unfriendly climates, while *A. m. ligustica* has been successfully introduced in warmer climates.

1.2.3. Transition zones, O And A Branch, and island populations.

The M and C branch are separated by the Alps, but separation is incomplete in the western coastal regions of Italy as well as in the eastern regions, where *A. m. mellifera* reaches south into the region of Belarus and meets the distribution area of *A. m. macedonica*. Settlers introduced this last race* into the Ukraine only about 500 years ago. Up to the northern regions of Ukraine, and southern of Belarus, some admixture with *A. m. caucasica* and *A. m. anatoliaca* can be found, which constitute the northwestern subspecies of the O branch bees. *A. m. caucasica* is the only bee of this branch which has become established in bee breeding on an international scale.

The O branch bees inhabit the Near East, they are smaller, light coloured, and characterised by wide metatarsi. *A. m. anatoliaca*, the bee of Turkey, is forming a transition zone with *A. m. macedonica* in the continental regions of Turkey.

A particularly strong introgression has taken place from the North African bees belonging to the A branch into Spain. From North to South, bees increasingly show characteristics of the bees of North Africa due to introgression of African genes. Though North African bees are morphometrically close, it is apparent from mtDNA investigation that these belong to the African branch, instead of the M branch (see below).

Local populations inhabit the Mediterranean Islands, some described as own subspecies. Characteristically, these combine influences from adjacent coasts, thus representing transitions with distinct own characteristics. The bees of Greek islands, including *A. m. adami* from Crete and *A. m. cypria* from Cyprus, belong to the O branch, which points to a predominant origin of these island populations from those of the Near East.

Likewise, the A branch inhabiting Africa has exerted some influence over the Mediterranean Islands. *A. m. sicula*, the bee of Sicily, though first allocated to the C branch on morphometric similarity, shows distinct characteristics of African bees, as multi-queen swarming and African molecular markers*. The recently described bee of Malta, *A. m. ruttneri* [25], is even more close to the North African *A. m. intermissa* than to the European bees. Some African influence is likely present in the bees of Corsica and perhaps Sardinia. The Balearic Islands bee populations, though clearly belonging to the M branch on morphometric grounds, show predominant molecular markers of North African populations [6], [9].

2. MOLECULAR LINEAGES OF *APIS MELLIFERA*

The characters of the external morphology used for establishing the races* of *Apis mellifera* are determined by many hereditary factors that interact between them and with the environment in a complex way. Therefore, the genetic basis of morphological characters is usually difficult to identify, as the effect of a particular gene is masked by the superimposed effect of many other genes also involved in determining a given character. However, at the molecular level there are genes available to racial and population studies that can be characterised univocally, as its nature is not masked by the action of other genes. Some of these genes can be sequenced or fragmented in pieces whose length can be separated by electrophoresis and measured. Others produce a protein that can be likewise analysed. These genes are known as **genetic (or molecular) markers***. Furthermore, these genes have enough variants some of which are geographically restricted, that is, they are associated to a race* or are present in a combination of different genes that is unique to a set of populations.

Genes that determine both morphological and molecular characters are part of the genetic pool of races and populations and thus, evolve intimately associated. For this reason molecular markers have become a powerful tool for racial studies. However, these markers show two main handicaps. One is that the way in which they change is not necessarily corresponded by the same changes in characters determining morphological traits. This means that a population showing a morphological homogeneity may have a large amount of hidden molecular variation and vice versa, a large amount of morphological variation may be associated to a molecular homogeneity in a given marker. A second handicap is that the study of the molecular markers is restricted to laboratories properly equipped and is not amenable to the custom beekeeper.

By the end of the eighties there was a fruitful coincidence of events that favoured a rapid development of molecular analyses of races and populations of *A. mellifera.*

The aforementioned book of Ruttner [23] greatly stimulated the search of new characters to attain a better definition of races of *A. mellifera* and the limits of their geographic distribution.

The need of detecting the rapid advance of Africanized bees in the Americas encouraged the development of accurate tests to assess the degree of introgression of invading populations. The papers of D.R. Smith pioneered this line of research (review in [26]).

In 1993, Crozier and Crozier [5] sequenced the whole mitochondrial chromosome* of *Apis mellifera*. This is a small and circular molecule made up of DNA that is found copied 5-10 times inside a cellular organ, the mitochondria*. Each cell of the bee has dozens or hundreds of these mitochondriae, and thus provides a good amount of hereditary material to be studied.

The development of the polymerase chain reaction* (PCR) in 1983 was followed by many new techniques for studying a wide variety of molecular markers, in a simpler, faster and cheaper way. The focus was set on the DNA, the hereditary material present in the nucleus and the mitochondriae of the cell.

The rapid increase of molecular studies on *Apis mellifera* can be easily assessed by examining any of the bibliographic lists available in Internet, as happens with http://www.geocities.com/BeesInd/index.htm. Yet, these studies will probably increase much more in the next future, thanks to the many possibilities that will arise from the project aimed to unravel the sequence of the *Apis* genome* (http://www.genome.gov/11008252 and http://hgsc.bcm.tmc.edu/projects/honeybee/).

The molecular markers are appropriate to study different problems. Some parts of the DNA change slowly with time and are thus helpful to reconstruct the events that happened some hundred thousands years ago. Other parts evolve at a high speed and are suitable for the study of recent events, of a few hundreds or thousands years only.

Not always but often, the genetic analyses performed on *Apis mellifera* have been based on previous information of the morphometric characteristics. In fact many of the published studies try to test the aforementioned hypotheses about the subspecies stated by Ruttner and colleagues.

The investigations based on molecular markers must take into account the particularities of the reproductive biology of the honey bee. There is normally one queen per colony whose unique function is to lay eggs. The queen controls whether the eggs are fecundated or not. In the first case they will develop into new queens and workers, and in the second into drones. This means that queens and workers have two copies of DNA (one inherited from the queen and other from the drone) in each cell of the body, a condition known as "diploid"*. On the other hand, the drone has a single copy of DNA* per cell, inherited from the queen. This condition is known as "haploid".

As the queen mates with up to 20 drones in her mating flights, in one colony there may coexist up to 20 worker subfamilies or patrilines*. Since all workers are offspring of the same mother queen, workers with the same father drone are related in the 75% of their hereditary patrimony (super-sisters) and those with different fathers are related in the 25 % (half-sisters). Thus, molecular markers are not evenly distributed in the colony members, a point that must be always taken into account.

We are going to focus on those widely used in studies about the subspecies of honey bees around the Mediterranean basin. Some others have being mainly used to investigate the africanization problem in honey bee populations from America and will not be addressed here.

2.1. Types of molecular markers

2.1.1. Mitochondrial DNA

Mitochondrial DNA* is found inside the mitochondria* (not inside the nucleus of cell) and therefore does not participate in sexual reproduction. It changes very slowly, only by mutation*. Mitochondrial DNA (mtDNA onwards) is transmitted only through the ovules (not through the sperm) and thus workers and drones have the same mtDNA coming from the queen. For this reason all members of the colony share a unique copy of mtDNA called the haplotype*. Likewise, mtDNA is used to trace maternal lineages. We will refer herein to the haplotype* distribution and characterisation as a way to discriminate European honey bee populations.

Cornuet and Garnery [4] began in the nineties with the analysis of the sequence of mitochondrial genes to discriminate among the *Apis* subspecies and to establish genetic relationships among them. They found 19 mtDNA types on a sample of ten different subspecies of *Apis mellifera* that clustered in three major phylogenetic lineages. These lineages corresponded well to three groups of populations with distinct geographical distributions: branch A for African subspecies (*monticola, scutellata, andansonii, capensis,* and *intermissa*), branch C for North Mediterranean subspecies (*caucasica, carnica* and *ligustica*) and branch M for the West European populations (*mellifera* subspecies). These results confirmed most conclusions based on morphometric characters and allozymes. However, the North African races, *A. m. intermissa* and *A. m. sahariensis*, show the same kind of haplotype* as the other African subspecies and for this reason molecular researchers tend to include them into branch A instead of branch M, as proposed by Ruttner in his classification of subspecies based on morphometry*.

Using a similar approach but with the information from other mitochondrial genes (part of NADH dehydrogenase subunit 2 and isoleucine transfer RNA genes), Arias and Sheppard [2] studied 14 subspecies *Apis mellifera* and the New World "Africanized" honey bees. Twenty different haplotypes were detected and phylogenetic analyses supported the existence of 3 or 4 major subspecies groups similar to those based on morphometric measurements. Minor discrepancies were reported concerning the subspecies composition of each group.

In 1993, Garnery and colleagues [17] developed a simple test to discriminate among mitochondrial haplotypes. They called it the *Dra*I test and consists on the amplification of a particular mitochondrial region that is lately fragmented with the restrictase *Dra*I. The resulting fragments are then separated by electrophoresis onto a gel according to their length. Each haplotype gives a characteristic pattern of bands easily observable (figure 2). Since the development of this technique numerous analyses have been carried out, mainly focussed on the genetic diversity of the west European populations. In concrete, Iberian, French and Italian honey bee populations have been submitted to extensive analysis of mitochondrial variation.

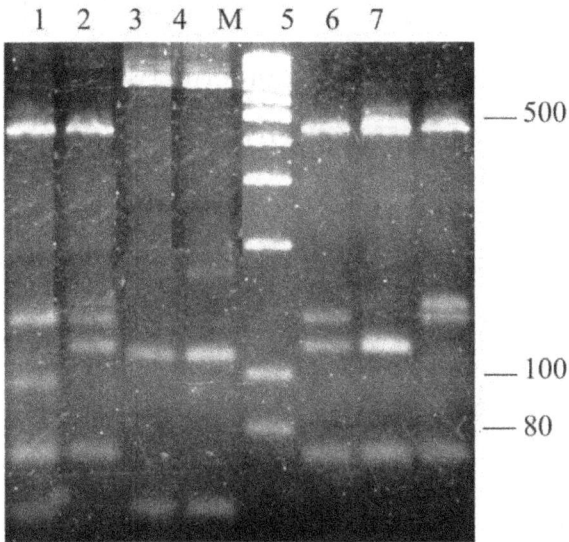

Figure 2 : **Dra**I **restriction pattern of seven *A. mellifera iberiensis* samples corresponding to the following haplotypes: A1 (lane 6) and A2 (lanes 3 and 4) belonging to the African lineage, M4 (lane 7), M7 (lane 1) and M8 (lanes 2 and 5) belonging to the Western European lineage. The 100 bp ladder is included as a marker (M).**

A. MELLIFERA IBERIENSIS: Honey bee populations of the Iberian Peninsula are of particular interest because of the hypothesised hybrid status of *A. m. iberiensis* between the European *A. m. mellifera* and the African *A. m. intermissa*. In agreement with this hypothesis, it was found that M west European haplotypes gradually decrease from north to south and the opposite happened with African (A) haplotypes [18], [19], [25]. The gradient is smooth on the east of the Iberian

33

Peninsula, as we have found in the BABE project after examining populations from Catalonia to Andalusia. In the province of Valencia, right in the middle of Spain, the proportion of M and A haplotypes is 50%. To the west the gradient is rather steep. Cánovas et al. [3] have shown that in Galicia (SW Spain) the two southernmost provinces (Orense and Pontevedra) show about 95% of African (A) haplotypes, and a few kilometres northwards the percentage of west European (M) haplotypes increases up to 96.9% (provinces of La Coruña and Lugo).

Preliminary results in other Iberian regions also corroborate the occurrence of the gradient of haplotypes. This suggests that it is a situation extended across the whole Iberian Peninsula. De la Rúa et al. (unpub. res.) have postulated that the gradient has a natural origin and reflects the colonising movements of both *A. m. mellifera* and *A. m. intermissa* populations. As noted by Ruttner [23] *A. m. mellifera* probably colonised west Europe after the last glaciation period (8000 –10,000 years B.P.) from refuges located in Iberia. This northward movement was possibly followed by *A. m. intermissa*, as soon as south Iberia became increasingly warmer. The mixing of genes and characters of both subspecies, either with adaptive or with neutral value, explain the hybrid nature of *A. m. iberiensis* and the distribution of different character sets (morphological, enzymatic, DNA, etc.).

A. MELLIFERA MELLIFERA: A wide survey of sixteen French localities was chosen to analyse the genetic diversity of the honey bees [19]. These populations showed the original haplotypes* of the western European honey bees that belong to the M evolutionary lineage*. French populations exhibit various levels of introgression from other evolutionary lineages, mainly the C lineage coming from *A. m. carnica* populations located in Germany. The level of introgression depends on the type of beekeeping techniques developed by the beekeepers: it is low in regions where amateurs are abundant and high where professional beekeepers regularly import foreign queens, mainly of *A. m. carnica*. The haplotypes M4 and M4' are the most frequent and widespread in west Europe (from Spain to Sweden) and perhaps represent a primitive condition.

A. MELLIFERA LIGUSTICA AND A. M. SICULA: The honey bees from Italy were morphometrically assigned to a single race, *A. m. ligustica*. However, mtDNA and microsatellite* analyses have clearly shown [13] that Italian honey bees are actually a mixture of M and C lineages over most of the Apennine Peninsula. This means that *A. mellifera ligustica* has a hybrid nature similar to that described for *A. m. iberiensis*. Franck et al. [13] concluded that bees of the M lineage (mostly M7) colonised first the western (Tyrrhenian) coast whereas the C lineage (haplotype C1) invaded Italy from the NE and followed the Adriatic coast. The likely isolation of Italian bees during Pleistocene climatic oscillations allows for estimating in 190.000 years the date of such colonisation events. A recent introgression of French populations (characterised by the M4 haplotype) in NW Italy has been also reported by these authors, a situation similar to that found in France with regard to *A. m. carnica* introgressing *A. m. mellifera*.

The hybrid origin of *A. m. ligustica* has long been obscured by the fact that in the main area of queen production (from which most of the previous ligustica bee samples originated) the M mitochondrial lineage is absent, whereas it is present almost everywhere else in Italy.

In Sicily the subspecies *A. m. sicula* shows an almost dominant presence of the African A2 haplotype (as in Southern Spain). M haplotypes have been found only in a few spots (remnants of ancestral colonisation?) by Sinacori et al.[28]. On the contrary, microsatellite* data (see above) show an Apenninian influence, that corroborates the mixture of influences in Sicilian populations. The fact that the A2 haplotype is also common in Ionian and Balearic islands, and in the Iberian Peninsula perhaps reflects an ancient colonisation from Africa (this haplotype has not yet been found in North Africa). Franck et al. [13] also consider the possibility of human dispersal of this haplotype in recent historic periods.

2. 1. 2. Microsatellites*

Microsatellites are sequences of nuclear DNA that consist of repeats of a simple sequence of nucleotides* (for example, AAT repeated 15 times in succession) situated one besides the other (tandem* arrangement) in the chromosomes* that make up the genetic material. Using the polymerase chain reaction* (PCR) these repeats can be easily amplified. There are many of these markers in a single individual and a given microsatellite usually shows many variants within populations, that is, they are extraordinarily variable. The number of repeat units that an individual has at a given locus can be easily resolved using polyacrlyamide gels or sequencing machines.

For a given microsatellite there are two marks in the gel per worker, as each individual inherits one nucleotide* repeat from the queen and another from the drone (a single mark would correspond to the same inheritance coming from the queen and the drone). Workers of the same hive may differ in a given microsatellite, as they share the same gene from the queen but may differ in the father. This peculiarity leads to one line of research, the identification of the patrilines*. After the study of many workers of the same colony (usually more than 50), it is possible to determine the genetic constitution or genotype* of the queen and of the different drones she mated with.

Microsatellite loci are extremely polymorphic* in African populations compared to European honey bee populations. This finding suggests that African populations have a larger effective size than European ones, that is, African populations probably have had a large and stable size since many thousands of years, whereas size has possibly fluctuated notably in Europe due to Pleistocene climatic oscillations (successive cycles of population growth and decimation).

Microsatellites were firstly used in studies about molecular evolution, theoretical models of mutation* and reproductive behaviour and sociobiology of the honey bee. More recently, a study over the paternity and maternity frequencies was performed on honey bee colonies of *Apis mellifera sicula*, the subspecies from Sicily [22]. They found that the queens living in this island mate with at least 5 to 12 drones and that the mating frequency may be positively correlated with drone density.

The studies on the variability of microsatellite loci usually include between from 7 to 11 of these markers. Estoup et al. [12] confirmed that *A. mellifera* evolved in the three distinct lineages –A, C, and M- previously detected by morphological and mtDNA analyses. Likewise, Franck et al. [14] corroborated the existence of the O lineage from the Near East after the analysis performed on a sample of honey bees from Lebanon. This O lineage was characterised by Ruttner [23] on morphological grounds and lately assessed by mtDNA analyses performed by Arias and Sheppard [2] and Palmer et al. (2000). The O lineage shows a marked genetic differentiation from neighbouring populations belonging to other evolutionary lineages*. A fifth lineage called Y [15], made up by populations of *A. m. jemenitica* present in Ethiopia, was defined according to particular mtDNA sequences and microsatellite composition.

Other aspects studied through the analysis of microsatellite loci have been the origin of West European subspecies of honey bees, their genetic diversity and the hybrid origin of the honey bees from Italy and Sicily. These nuclear markers brought about complementary information that derived from mtDNA analyses, giving usually a congruent -and improved- picture about honey bee evolution. As discussed below, differences in the inheritance mechanism of mtDNA and microsatellites cause sometimes apparently incongruent results, but these problems should be eventually solved with new data from different sources.

The genetic profiles of the western European populations have been shown to be rather homogeneous from Spain to Scandinavia, although some controversy exists in relation to the Iberian honey bee. Microsatellite data of forty-five unmanaged honey bee colonies from the Southeast of the Iberian Peninsula [8] relate them to the African *A. m. intermissa*, although some alleles* and the observed heterozygosity* were characteristic of European *A. m. mellifera* populations. These results corroborate the hybrid nature of *A. m.*

iberiensis. Interestingly, no recent introgression* from Africa was detected, which does not support human influences in historic times.

Microsatellite analyses of French populations [20] corroborate that introgression* from bees of the C lineage (mostly from *A. m. carnica* and *A. m. ligustica*) is occurring at various localities, in close correspondence with a high presence of mtDNA of that lineage. In Angers, Versailles, Orsay and Fleckenstein, professional beekeepers have imported many foreign bees to replace missing colonies infested by *Varroa*, which has caused a dramatic change in the gene pool of local populations. This situation resembles the one experimented in German populations 55 years ago when native *A. m. mellifera* populations were all replaced by *A. m. carnica*.

3. CORRELATION BETWEEN MORPHOMETRIC AND MOLECULAR CHARACTERS.

Ruttner [23] considered and summarised the whole evidence to define the subspecies of *Apis mellifera*. Morphometry*, behaviour, and ecological characters were taken into account for this purpose. As noted above, his classification has been the basis for late studies based on molecular characters. It has been found there exists a good correlation between Ruttner's proposals and molecular conclusions, as allozymes, mtDNA and microsatellites* have corroborated the four main branches of subspecies (A, C, M, and O). Likewise, many subspecies included within each of these branches have been also characterised by detailed molecular traits. However, up to now a complete subspecies description based on molecular markers is not available and morphometry is still the main tool to determine and define subspecies membership, but there is little doubt that this is under way and will emerge in near future.

A close correspondence between morphometric characterization and molecular markers is not surprising, as some of the measured morphometric traits are apparently fairly conservative over evolutionary time and inert to environmental influences. However, others appear to adapt more rapidly to specific local conditions. Diniz-Filho *et al.* [10] assessed the relative influence of phylogeny* as represented by the mtDNA pattern and that of the geography (predominantly latitude) on the various traits. They showed that particularly the measurements of wing venation patterns were predominantly under control of phylogeny and body size correlated more to latitude, while others as pigmentation varied locally without strong correlation to these large-scale determinants. Morphometric characters thus differ in their usefulness when attempting to characterise the origin of races* and to reconstruct the evolutionary biology of European honey bees. Some of these characters respond strongly to local conditions (e.g., those related to body size) and thus their content of historic information is low, while others, e.g. wing venation evolve over several thousands or millions years without an apparent environmental influence, so their history tells also the history of the races and subspecies. Combined analysis together with molecular data can be extremely insightful to unravel the interrelations of historical events and local selection pressures on the morphological features of honey bees.

Molecular data have also added a better understanding of the genetic architecture and evolutionary history underlying the variation showed by morphological subspecies. Thus, *A. m. iberiensis* is geographically intermediate between *A. m. intermissa* and *A. m. mellifera*, and shows morphological behavioural traits of these two subspecies [23]. Its presumptive hybrid nature has been fully corroborated by molecular data, which show a strong and geographically wide introgression* between the two later subspecies across the Iberian Peninsula.

The fruitful interaction between these different data sets is the likely trend for incoming research in biogeography of honey bees and racial differentiation. Recently, two new subspecies have been described on the basis of morphometric analyses (*A. m. ruttneri* from Malta [25] and *A. m. pomonella* from Central Asia, [27], and some molecular characters (sequence analysis of mtDNA) were included in the definition.

Molecular data have thrown into light some problems not previously envisaged with traditional systematics*. Three of these have been briefly mentioned above. The first is that *A. m. intermissa* is a member of the western European branch (M) in Ruttner's view

[23], but shows mtDNA and microsatellite* characters typical of African subspecies (A lineage). The second is the hybrid nature of *A. m. ligustica*, which shows a good racial characterisation and harbour genetic influences of both M and C lineages. Thirdly, honey bee populations from Ethiopia are members of the A branch (*A. m. jemenitica*) but are quite distinctive in mtDNA and microsatellites*, this has lead to Franck et al. [15] to create a fifth molecular lineage called Y for including this subspecies.

These relative disagreements merely tell us that we have not yet enough data to understand all the factors and forces responsible for the genetic basis underlying the racial traits of honey bee populations. These factors include not only present selective forces and recent human influences but also the evolutionary events that exerted a deep influence in colonisation success, population extinction, trends in adaptation, etc. As soon as new data become available these disagreements will eventually disappear.

Some disagreements are even found between mtDNA and microsatellite* data. These are due to the different transmission mechanism of these markers. A hive founded by an Italian queen imported in South Spain will show a C haplotype (mtDNA) during successive generations, as her daughters receive this haplotype only via maternal inheritance. This marker will be detectable until the colony is physically removed or the queen dies. However, morphological and behavioural traits, microsatellites*, and any other character controlled by the hereditary material located inside the cell nucleus will be progressively diluted, as local drones contribute with 50% of the genetic material of new queens in each reproductive season. Therefore, one would find an indication of queen importation in the mtDNA but this might not be corresponded in the microsatellite* analysis.

According to this explanation races geographically intermediate between Africa and Europe may show a high proportion of African haplotypes (introgressed due to the higher dispersal power of African populations), but are more "European" in morphological traits and microsatellites. It is also possible that mtDNA haplotypes are subjected to particular selective forces, that might influence the settlement or elimination of alien queens bearing a different haplotype. This hypothesis has given rise to a model that is being tested for the Iberian Peninsula.

4. CONCLUDING REMARKS

1. European races* of *Apis mellifera* are members of M and C branches of Ruttner's classification [23]. Races between Europe and Africa show a varying degree of introgression* from African subspecies, mostly *A. m. intermissa*. This is the case of *A. m. iberiensis, A. m. siciliana*, and –particularly- of *A. m. ruttneri*. The introgression* is more evident in the haplotypes corresponding to

mitochondrial DNA, and is lower in morphological traits and microsatellites*.

2. European races of *Apis mellifera* are the result of different colonising events. According to the hypothesis of Garnery et al.[16], a number of lineages were originated in the Near East about one million years ago. Of these, the M lineage reached Western Europe possibly by crossing central Europe, whereas bees of the C lineage colonised the Balkans and entered in east Italy. Lately, Mediterranean islands, south Iberian Peninsula and south Italy have been possibly populated one or more times by bees belonging to the A lineage (figure. 3).

3. Populations of the M lineage have probably undergone dramatic size changes due to climatic oscillations during the last thousands million years. Native honey bee populations from Iberia to Scandinavia are perhaps the descent of populations that found shelter in Iberia (and perhaps in south France) during the last glaciation period. A hypothetical oriental refugee has been also suggested by Garnery et al.[19]. Though populations of the M7 lineage also occurred in Italian refugees, they have not probably participated in the re-colonisation process and stayed isolated. The occurrence of these M7 haplotypes also in Spain [13] indicates a recent natural or man-made interchange between Italy and Spain.

4. Italian bees are the result of two waves of colonisation. The M lineage probably entered the Apennine Peninsula by the western side

about 190.000 years ago whereas the C lineage (perhaps a little bit later) colonised initially the eastern side. However, selection and exportation of *A. mellifera ligustica* has been concentrated on bees having the C haplotype and therefore the distribution of this marker in the Apennine Peninsula and other countries and islands has usually a strong anthropogenic origin.

5. Further research is badly needed in the subspecies of the C lineage occurring in SE Europe. Morphology alone is not enough to characterise populations and assure a full assignment. The combination of mtDNA and microsatellite* analyses may improve their racial characterisation and help in reconstructing the evolutionary events (dispersal routes and colonisation, extinction, recoveries, etc.) of honey bee populations in this area.

6. The geographic distribution of European honey bee races, as well as their genetic background, may be dramatically changed in the next years due to beekeeping practices. Large-scale replacement of native populations in search of 'suitable' characteristics, massive importations associated to severe decimation of populations by *Varroa* and other parasites, and migratory beekeeping are processes that have already shown their powerful effects. To the examples mentioned above, it should be added the marked genetic introgression* detected in some Balearic [6] and Canarian Islands [7]. As the problem has an international dimension, UE authorities should eventually take decisions to make compatible a full

development of European beekeeping and the maintenance of native gene pools as free of introgressed genes as possible.

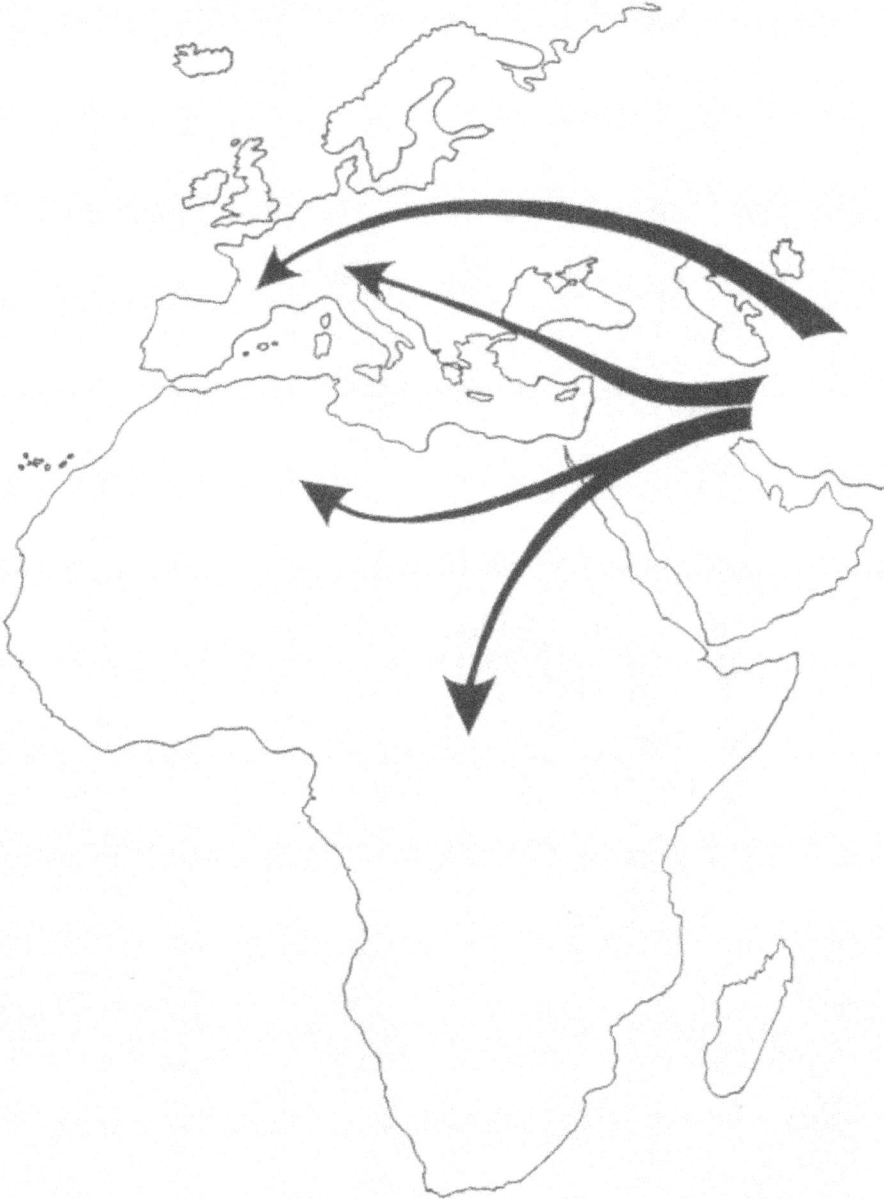

Figure 3 : **Hypothesis about the origin and dispersal of the honey bee based on distribution of mitochondrial DNA lineages. Slightly redrawn from Garnery et al. [16].**

References

[1]   Alpatov W.W. 1929. Biometrical studies on variation and races of the honeybee *Apis mellifera* L. Quaternary Review of Biology 4: 1-57.

[2]   Arias M.C., Sheppard W.S. 1996. Molecular phylogenetics of honey bee subspecies (*Apis mellifera* L) inferred from mitochondrial DNA sequence. Molecular Phylogenetics and Evolution 5 (3): 557-566.

[3]   Cánovas F., De la Rúa P., Serrano J., Galián J. 2002. Variabilidad del ADN mitocondrial en poblaciones de *Apis mellifera iberica* de Galicia (NW España). Archivos de Zootecnia 51: 441-448.

[4]   Cornuet J.-M., Garnery L. 1991. Mitochondrial-DNA variability in honeybees and its phylogeographic implications. Apidologie 22 (6): 627-642.

[5]   Crozier R. H., Crozier Y. C. 1993. The mitochondrial genome of the honeybee *Apis mellifera*: complete sequence and genome organization. Genetics 133: 97-117.

[6]   De la Rúa P., Galián J., Serrano J., Moritz R.F.A. 2001a. Molecular characterization and population structure of the honeybees from the Balearic Islands. Apidologie 32: 417-427.

[7]   De la Rúa P., Galián J., Serrano J., Moritz R.F.A. 2001b. Genetic structure and distinctness of *Apis mellifera* L. populations from the Canary Islands. Mol. Ecol. 10:1733-1742.

[8]   De la Rúa P., Galián J., Serrano J., Moritz R.F.A. 2002. Microsatellite analysis of non-migratory colonies of *Apis mellifera iberica* from south-eastern Spain. Journal of Zoological Systematics and Evolutionary Research 40: 1-5.

[9]   De la Rúa P., Galián J., Serrano J., Moritz R.F.A. 2003. Genetic structure of Balearic honeybee populations based on microsatellite variation. Genetics Selection Evolution 35:339-350.

[10]   Diniz-Filho J.A.F., Fuchs S., Arias M.A. 1999. Phylogeographical autocorrelation of phenotypic evolution in honey bees (*Apis mellifera* L.). Heredity 83: 671-680.

[11]   Engel M.S. 1999. The taxonomy of recent and fossil honey bees (Hymenoptera: Apidae; Apis). Journal of Hymenoptera Research 8: 165-196.

[12]   Estoup A., Garnery L., Solignac M., Cornuet J.-M. 1995. Microsatellite variation in honey-bee (*Apis mellifera* L.) populations- hierarchical genetic-structure and test of the

infinite allele and stepwise mutation models. Genetics 140 (2): 679-695.

[13] Franck P., Garnery L., Celebrano G, Solignac M., Cornuet J.-M. 2000a. Hybrid origins of honeybees from Italy (*Apis mellifera ligustica*) and Sicily (*A. m. sicula*). Molecular Ecology 9 (7): 907-921.

[14] Franck P., Garnery L., Solignac M., Cornuet J.-M. 2000b. Molecular confirmation of a fourth lineage in honeybees from the Near East. Apidologie 31 (2): 167-180.

[15] Franck P., Garnery L., Loiseau A., Hepburn H.R., Solignac M., Cornuet J.-M. 2001. Genetic diversity of the honeybee in Africa: microsatellite and mitochondrial data. Heredity 86:420-430.

[16] Garnery L., Cornuet J.-M., Solignac M., 1992. Evolutionary history of the honey bee *Apis mellifera* inferred from mitochondrial DNA analysis. Molecular Ecology 1(3): 145-154.

[17] Garnery L., Solignac M., Celebrano G., Cornuet J.-M. 1993. A simple test using restricted PCR-amplified mitochondrial DNA to study the genetic structures of *Apis mellifera* L. Experientia 49:1016-1021.

[18] Garnery L., Mosshine E.H., Cornuet J.-M. 1995. Mitochondrial DNA variation in Moroccan and Spanish honey bee populations. Molecular Ecology 4:465-471.

[19] Garnery L., Franck P., Baudry E., Vautrin D., Cornuet J.-M., Solignac M. 1998a. Genetic biodiversity of the West European honeybee (*Apis mellifera mellifera* and *A. m. iberica*). I. Mitochondrial DNA. Genetics Selection Evolution 30: 31-47.

[20] Garnery L., Franck P., Baudry E., Vautrin D., Cornuet J.-M., Solignac M. 1998b. Genetic biodiversity of the West European honeybee (*Apis mellifera mellifera* and *A. m. iberica*). II. Microsatellite loci. Genetics Selection Evolution 30: 49-74.

[21] Götze G. 1964. *Die Honigbiene in natürlicher und künstlicher Zuchtauslese*. Paul Parey, Hamburg

[22] Haberl M., Tautz D. 1999. Paternity and maternity frecuencies in *Apis mellifera sicula*. Insectes Sociaux 46: 137-145.

[23] Ruttner F. 1988. *Biogeography and Taxonomy of Honeybees*. Springer, Heidelberg.

[24] Ruttner F., Tassencourt L., Louveaux J. 1978. Biometrical-statistical analysis of the geographic variability of *Apis mellifera* L. Apidologie 9: 363-381.

[25] Sheppard W.S., Arias M.C., Meixner M.D., Grech A. 1997. *Apis mellifera ruttneri*, a new honey bee subspecies from Malta. Apidologie 28 (5): 287-293.

[26] Sheppard W.S., Smith D.R. 2000. Identification of African-derived bees in the Americas: A survey of methods. Annals of the Entomological Society of America 93 (2): 159-176

[27] Sheppard W.S., Meixner M.D. 2003. *Apis mellifera pomonella*, a new honey bee subspecies from Central Asia. Apidologie 34 (4): 367-375.

[28] Sinacori A., Rinderer T.E., Lancaster V., Sheppard W.S. 1998. A morphological and mitochondrial assessment of *Apis mellifera* from Palermo, Italy. Apidologie 29: 481-490.

CHAPTER 2

SELECTION THEORY AND EFFECTIVE POPULATION SIZE

Michel Solignac, Jean-Marie Cornuet

1. GENERAL THEORY

Artificial selection, as opposed to natural selection, is the result of a breeding program performed by human. The purpose is generally to change the frequency of a qualitative character or the average value of a quantitative character in a given population. This is performed by breeding together individuals (animals or plants) that have been "selected". It seems evident that this will work only if the character has some (at least partial) genetic determinism. If the determinism is due to a single gene, there is no need for a sophisticated theory, but if it is due to many genes, a theory has been developed to help select individuals and organize their crosses in the most efficient way. As a matter of fact, most of characters of interest for breeders are quantitative and are determined by at least several genes*. Quantitative genetics is the part of the genetic science that deals with quantitative characters and its main output is the selection theory (see [7] for a general introduction).

1.1 Decomposition of the phenotype

Consider a quantitative character (e.g. the body size). The observed value for a given individual is called the individual's *phenotype** (*P*). The phenotype can be decomposed into a character value due to the particular assemblage of genes possessed by the individual and a value which encompasses all non-genetic effects. The first value is called the *genotype** (*G*) and the second is called the *environment* (*E*). So we have the classic equation $P = G + E$.

The *genotype* can be further decomposed in an *additive value* (*A*), a *dominance deviation* (*D*) and an *interaction deviation* (*I*) : $G = A + D + I$. The additive value and dominance deviation are the sum of additive values and dominance deviations of all loci acting on the character, respectively. The *additive value* is also termed the *breeding value*. It is the expected value of the progeny of the individual. It is the fraction of the genetic value that can be transmitted to the progeny. It is then of primary interest for the breeder to have a correct idea of the fraction of the variation of phenotypes that is due to the variation of the breeding value. The dominance and interaction deviations are the residual values (when the breeding value is removed from the genotypic value) when considering loci* separately and together respectively.

1.2 Heritability and response to selection

Consider an experiment in which a quantitative character has been measured on 100 offspring and their two parents. The results can be represented by points in a two-dimensional graph (fig. 1), the coordinates being the mean phenotypic value of parents (=mid-parent value) and the offspring value respectively. If offspring tend to resemble to its parents, (i.e. parents with larger values tend to have offspring with larger values), there will be a parent-offspring correlation that can be materialized by a regression line with a positive slope. The steeper the slope, the higher the correlation (the resemblance). In the particular case of the mid-parent-offspring relationship, the slope of the regression is precisely equal to the heritability* (h^2), which is the ratio of the phenotype variation that is due to the variation of breeding values of individuals in a population. In more precise terms, it is the ratio of the additive variance over the phenotypic variance. The additive variance is simply the variance of additive (= breeding) values.

Suppose now that, in the same experiment, the 15 pairs of parents with higher value are selected. Let S be the selection differential, i.e. the difference between the average of these 15 pairs of individuals and the average of all the parents. The selection response R, i.e. the difference between the average of the selected parents' offspring and the average of all offspring, can be estimated through the regression analysis by $R = bS$, where b is the slope of the regression line which is equal to the heritability*. So we have the following basic relationship :

$$R = h^2 S$$

In other words, the response to selection in a basic individual selection scheme is proportional to the heritability of the character and to the selection differential. The breeder can modulate the selection differential through the selection pressure (i.e. the proportion of selected individuals). He can also try to increase the heritability by enlarging the genetic basis of his population under selection. However, this can be done essentially at the beginning of a selection scheme. Note also that the term heritability is somehow confusing. A character with a strong genetic determinism can have a very small heritability when most individuals have the same alleles in the population. Also note that different populations may have different values of heritability for the same character.

Because of its impact on the success of selection, the heritability is a parameter that is worth being estimated before (and even during) any selection trial. Offspring-mid-parent regression is only one among various alternatives (e.g. sib analysis or intra-sire regression of offspring on dam).

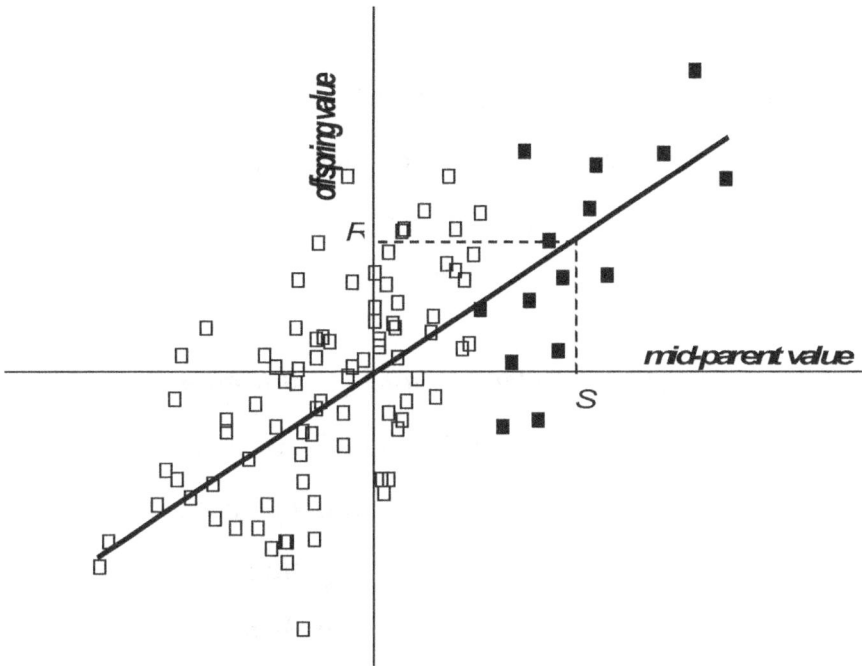

Figure 1 : Relationship between the mid-parent value and the offspring value. The thick black line represents the regression line for all observations. Filled squares correspond to selected individuals, S being their average value. R is the expected response to selection.

1.3 Index selection

The individuals to be selected are those with the "best" breeding values. It is then crucial to have the best estimate of an individual's breeding value. This can be achieved by combining various sources of information, such as its own phenotype or the phenotypes of related individuals. Usually, the combination of these sources of information is performed through a multiple regression analysis and the index of the individual is the linear combination of self and relative phenotypes that maximizes the correlation between the index and the breeding value of the individual. Index selection is more efficient than simple individual selection. It is necessary in some selection schemes such as combined selection, or when the character can not be measured on selection candidates (e.g. milk production in bulls).

1.4 Marker Assisted Selection and Quantitative Trait Loci

When a character is difficult or costly to measure or when it is measurable in only one sex, then it can be more advantageous to select for a secondary character that is genetically correlated to the first one. This is termed *indirect selection*. *Marker Assisted Selection* (*MAS*) is a special case of this indirect selection, in which secondary characters are DNA* markers (such as microsatellites*). The advantage of indirect selection occurs when the product of the genetic correlation between the primary and secondary characters by the heritability of the secondary character is larger than the heritability of the primary character. Polymorphic DNA markers have high heritabilities and can be taken in numbers such that some of them can be highly correlated with the genes involved in the determinism of the primary character.

Most of the quantitative genetics theory relies on the assumption that many genes with small individual effects are responsible of quantitative characters. However, it appears in practice that many characters are actually under the determinism of a small number of genes with some of them having a major effect. With a large set of

DNA markers which have been mapped on chromosomes*, it is possible to determine loci* (short chromosome fragments) involved in quantitative characters, called *QTLs* (*Quantitative Trait Loci*). Once this has been done, *MAS* based on markers close to these *QTLs*, can become much more effective.

2. SELECTION THEORY FOR HONEYBEES

2.1 Why selection theory cannot simply be applied to honeybees

Selection theory has been developed for species that can be bred under human control. In honeybees, natural fecundation occurs outside of the hive, generally in drone congregation areas and control on matings is very reduced unless with artificial insemination. The latter exists since the middle of the last century, however it requires a high level of technical practice and even the best technicians do not have a 100% success. It follows that honeybee breeders prefer selection schemes that can be achieved essentially with natural fecundation.

A second difficulty arises from the polygamy of queens. Queens may be inseminated by 5 to 30 males (see chapter on Effective population size). Two daughters of the same queens can be either full-sisters (same father) or half-sisters (different fathers). This uncertainty about their exact relationship implies an uncertainty in the computations of selection indices, and consequently in the best choice of candidates.

It may be desirable to select for worker characters, such as their defensive behavior or their ability to resist to parasites. Workers being sterile, one has to adapt formulae and selection schemes to take care of the true relationships between individuals. The usual mother-daughter relationship has to be replaced by an aunt-niece relationship (which is not defined precisely, as stated above).

Eventually, the characters of economic importance generally refer to the whole hive (e.g. the honey production) which is made of two successive generations, the queen (mother) and the workers (daughters). Three solutions have been proposed : i) consider the colony performance as that of an average worker, ii) consider that the colony performance is the queen's performance and iii) a synthesis of both previous solutions.

2.2 Decomposition of the performance of a colony

We will describe here the third solution, because it appears to be the most complete and to provide the best model to date for a character measured on a whole colony. The phenotype of the colony (P) is considered as the sum of three terms : a genetic contribution (Q) of the queen, an average genetic contribution ($\bar{W}$) of workers and a residual contribution (E) that includes all non genetic effects :

Computations that will not be presented here (see [4, 5] for details) require the definition of several other parameters :

-additive and dominance variances for the queen and the worker contribution respectively,

-additive and dominance covariances between queen and worker contributions,

-average kinship and identity coefficients of workers, which are directly related to the average number of drones mated to the queen.

This model shows that there is no clear definition of the heritability of a colony-level character. However, this definition is not necessary to compute selection indices and to estimate the response to selection (see below). Also it is possible to estimate the different components of the variance. This has been performed by Bienefeld and Pirchner [3] who analyzed performances gathered over 5581 colonies. These concern 5 traits of major economic importance among which are the honey yield, the wax production and the defensive behavior. A very interesting finding is the high negative genetic correlation between queen and worker contributions for all 5 characters. In other words, a fraction of the genes that make "good" queens also make "bad" workers (relatively to the trait under selection). This is understandable : considering that queens and workers have the same genes but have opposed physiological and behavioral characters, genes involved in these characters may have contrasting effects on both casts. This means that in the population under study, selection will have a limited success because improving one cast contribution will deteriorate the other cast contribution.

2.3 Index selection and expected response for a colony character

As mentioned above, a selection index is an estimator of the breeding value of an individual. Transposing it for a colony, we define the index as an estimator of the performance of the daughter colony, i.e. the colony in which the queen is a daughter of the queen of the considered colony. The information used by the selection index varies along with the selection scheme. Formulae have been given for a combined selection scheme in which colonies are selected according to their own performance and to the performances of "sister" colonies, i.e. colonies with sister queens (see [6] for details). The response to this combined selection has been compared with that of a mass selection. The most favorable situation for combined selection is when the family size is large and the genetic variances are low.

2.4 Inbreeding and sex locus

Inbreeding results in modified behaviors of workers : the thermoregulation of the nest and the recruitment activity are deteriorated, rearing and cleaning activities are reduced leading to an increased sensitivity to parasites. These examples confirm the extreme sensitivity of honeybees to inbreeding. Note that the multiple mating of queens in congregation areas full of males from many surrounding colonies is an efficient mechanism against inbreeding. Moreover, the sex locus* contributes to the elimination of a fraction of crosses between relatives. Both mechanisms tend to limit inbreeding in natural honeybee populations.

By reducing the number of progenitors, selection inevitably provokes an increase of the inbreeding level of populations. When the character is sensitive to inbreeding depression (most of characters of economic interest are, as explained above), part of the benefit due to selection will be lost. It is then crucial to evaluate the rate of increase of the inbreeding level when devising a selection scheme. This has been done for the combined selection scheme cited above. Computations have shown the interest of introducing immigrants in the selected population which can easily be done by allowing the natural fecundation of queens. This will reduce slightly the expected response to selection but the inbreeding level will increase at a much lower rate.

The maintenance of a sufficient number of sex alleles is also of concern when studying any selection scheme. Individuals that are homozygous* to this locus* are eliminated during their larval life, producing "holes" in the brood. When the number of alleles is reduced, the probability of homozygotes increases and the average brood viability is reduced [13] leading to a general decrease of economic traits. Computing the expected number of sex alleles along the generations of a selection scheme is difficult. However, precious indications have been obtained by computer simulations [11]. For instance, starting with 12 sex alleles they found a probability equal to 0.58 that there remain at least 6 of them after 10 generations if 15 queens are selected at each generation.

2.5 Molecular tools and the future of honeybee selection

Two classical categories of molecular markers have been developed in honeybees : RAPDs and microsatellites*. Both have been used to build genetic maps [12]. Thus we dispose of a large number of mapped DNA markers allowing to determine QTLs. This has already been achieved for several characters such as defensive and foraging behaviors [9, 10].

In parallel, the sex locus has been cloned* and several of its alleles have been sequenced [2, 8] allowing the direct identification of individual bee genotypes. Eventually, the complete DNA sequence of the honeybee genome will be soon available.

Classical selection theory which has been largely successful in many agricultural species did not bring so much progress in honeybees, partly for theoretical reasons as summarized above and partly for intrinsic reasons such as negative genetic correlations between queen and worker contributions. The recent and thorough development of molecular tools provides new exciting possibilities for genetically improving performances of honeybee colonies.

3. EFFECTIVE POPULATION SIZE

3.1 Generalities

The genetic variability of populations depend on a small number of factors :

1) The size of the population which determines the number of gene copies ; this number is the same for all genes, except some peculiarities.

2) The mutation* rate, or more exactly the rate of mutations which are not too deleterious and may persist in the population ; this rate varies from one gene to another one.

3) The selective forces which act on most genes ; several mode of selection exist resulting either in the elimination of deleterious alleles / fixation of advantageous ones or responsible for a polymorphic equilibrium for two or more alleles (produced by the advantage of heterozygote, or in some cases the advantage of rare alleles, of one allele in one sex and of the other allele in the other sex…).

4) The migration rate, individuals from neighbouring populations or imported from remote countries may carry alleles which do not exist in the population; their introduction modifies its genetic profile.

These factors, taken alone, are relatively easy to measure experimentally. However, they all play a role in the polymorphism and they may have synergetic or antagonistic effects. Moreover, the variability is also dependent on the history of the population, of the past population dynamics, even in very remote times, as well as past hybridisation events.

In the preceding list, we will mainly consider the population size. Several methods are available to determine this size, the most known one using the principle of dilution, the so-called « capture – recapture » method. It suffices to determine the percentage of labelled individuals in the recapture to get an estimation of the number of individuals in the area studied. Note however that this area is not necessarily a population. Even if the possibility of homogenisation of individuals before recapture is a prerequisite for the correctness of calculations, it does not necessarily coincide with the area occupied by interbreeding individuals.

This ecological estimation is however a very rough approximation because it is generally not in direct relationship with the polymorphism of the population. The polymorphism is in fact dependent on the so-called *effective* size of the population. This is an abstraction defined as the size of an *ideal* population that would have the same polymorphism than that observed in the *real* population under consideration. An ideal population is isolated (non-migration), its size is the same at every generation and individuals mate randomly. They have all, whatever their genetic constitution, the same fecundity and survival (no selection).

To approach a more realistic effective population size from the observed (ecological) size, some additional considerations are necessary. The first one is to consider only reproducing individuals (it is obviously rarely known). For various reasons, numerous individuals do not participate to the reproduction and their genes are lost for the next generation. It depends also of the ratio between males and females. If their numbers are not the same, the effective population size is not the sum of individuals of both sexes but is closer to the less numerous sex. This may occur in nature were a few dominant males inseminate most of the females or in agriculture with instrumental insemination (for example in bovines). Similarly, the variation of the number of descendants per progeny is important : more heterogeneous it is and more sensitive is the effect. Finally, it depends on the variations of the size of the population (over seasons or over years) and again the effective population size is close to the lowest values. As a consequence, the effective population size is always lower or far lower than the number of individuals observed or calculated with an ecological method.

From a genetic point of view, the very important phenomenon linked to the population size is called genetic drift (note that this drift has no relation with apicultural drift). The results of genetic drift on genetic variability is easier to understand with neutral genes (i. e. the different alleles of a gene have no noticeable effects on the survival or the reproduction of individuals or, in other words, whatever their genotype, the individuals are equivalent). Even for neutral alleles, the progeny do not exactly receive a gene pool strictly identical to that of the general population or even its reproducers. For instance, a mother with two copies of a gene that has two children may transmit the same copy to both (with a probability of 1/2) and the second copy is lost. A mother who has two boys transmit its mitochondrial DNA to both but because boys do not transmit this cytoplasmic molecule, her copy of mtDNA is lost in future generations. In this simple case where the size of the population and hence the number of gene copies is assumed to be the same at every generation, one can expect that if some copies disappear, other will be transmitted twice.

In the course of generations, the genetic drift has two consequences : i) mutations that are very rare may disappear very easily and rapidly ; ii) the frequency of variants which are not too rare will change at every generation. This process is purely random, like a brownian movement, and it will arrive that when a variant has greatly increased in frequency, it may be fixed and those which have greatly decreased in frequency may be lost. When the population is very small, this process is very rapid but it takes a very long time in large populations. Whatever the population size, if this process was the only one in action, all populations will lost their variability. This will simply take a longer or a shorter time. For mtDNA, this time, expressed in generations, is on average twice the number of females (1,000 generations for 500 females) and, for the genes of the nucleus (two copies in each sex), it is four times the number of individuals (4,000 generations for 1,000 reproducers).

When the process is achieved, the population is comprised of individuals homozygote for all the genes and all individuals from the population have exactly the same gene copies. Even if the population reproduces sexually, it is genetically equivalent to a clone. The sensitivity to a pathogen is the same for all individuals and genetic adaptation to a new environment is impossible.

The factor which counterbalances the loss of variability through genetic drift is the occurrence of new variants through mutation*. The role of mutation is generally underestimated because it is very rare to observe a new mutation. As a matter of fact, they are rare for a single gene in the interval of a single generation. However, as aforementioned, the genetic drift expends on a more or less large number of generations and during that time new mutations are the rule. Mutation and drift have opposite effects : drift erases genetic variability and on the contrary, mutation is at the origin of new variants.

In that case, an equilibrium is established when the variability lost by genetic drift is exactly compensated by the appearance of new variants by mutation. This mutation-drift equilibrium is stable as long as the population size and mutation rate is maintained. However, it has to be noticed that the new mutations do not restore the same variants than those that have disappeared by drift. Most of the variants which appear by mutation are totally new. This means that they never previously existed before in the population. In that respect, the polymorphism is quantitatively maintained in the course generations but it changes qualitatively because the nature of its variants is changing. In other words, the population evolves with time, modifying its polymorphism, which however keeps a constant level.

This mutation-drift equilibrium depends only on two factors : the effective population size, which regulates the genetic drift and the mutation rate.

It must be kept in mind that this model is an oversimplification because only two evolutionary forces are taken into account. It may be directly applicable to the region of the genome which have no effects or very low effects on the fitness of individuals, i. e. the less important regions of the genome, such as the large regions which surround genes and have no known functions, some regions of the genome which have an identified function but for which a precise sequence is not very important or, within the genes themselves, the mutations which do not change the structure of the product of the gene (the sequence of the protein), mainly the introns and a large part of mutations occurring on the third bases of codons. If only this « neutral space » of the genome is considered, and assuming that the mutation rate is rather uniform, because the number of copies is the same for all these sequences, the polymorphism is also rather uniform. It has to be noticed that some « exotic » objects of the genome such as microsatellites (repeats in tandem of two or more bases) have a particular mode of mutation (change of the number of repeats) which is far higher than the base substitution rate and hence they have a higher polymorphism. For mitochondrial DNA, the effective population size is lower (see above) but the mutation rate is far higher and the polymorphism is important, particularly in the control region deprived of genes.

For the coding part of the genes which are under selection, the spontaneous base substitution rate is similar to that of the neutral regions. However, at the difference of the latter, because some mutations are not compatible with survival or reproduction, the individuals which carry these mutations are rapidly eliminated and the mutation with them (negative selection). The consequence is that only a part of mutations (the non-deleterious part) participates to the mutation – drift equilibrium and consequently the polymorphism is lower. More the sequence of a gene is under a strong constraint, and higher is the proportion of deleterious mutations. Consequently, the polymorphism of the genes is generally the reflect of their constraint : the higher is the constraint, the lower is the polymorphism. Another type of selection may occur (when the heterozygote has an advantage). This is the way to preserve polymorphism and it will be considered below.

However, even if we have reached a noticeable complexity to account for the value of the polymorphism, the picture is not still complete. The past demographic events of the populations have a strong effect on the current level of polymorphism and there are several situations. The changes in population size may be very important in case of climatic changes, new parasitosis and so on. In that case, the population size may be divided by ten or hundred in the interval of one generation (bottleneck) ; but, if the cause ceases rapidly, because many organisms have a high fecundity, the population size may be restored in a few generations. In that case, the polymorphism is practically maintained at the level he had before the bottleneck. However, when the population size decreases drastically and for a long time, the polymorphism drops down because the genetic drift is very strong in very small populations. In that case, when the population size is restored, the polymorphism keeps the level he had during the small population episode : only the variants preserved during the bottleneck are present, each having simply a higher number of copies. The population which has not the level of polymorphism expected from its current size is said to be in disequilibrium. A very high number of generations is needed to restore the polymorphism corresponding to the population size. A well-known example is the human species counting several billions of individuals and which has a polymorphism corresponding to only 10,000 individuals, the size of humanity 5,000 generations ago. The human beings have conserved the polymorphism of the original population, which has not followed the rapid demographic expansion.

4. POPULATION SIZE OF THE HONEYBEE

These factors described above are valuable for all species but the honeybee presents some peculiarities.

For many people, the honeybee is one of the most familiar animals in our countries and a simple calculation suggests a number of many billions of individuals in several European countries. In fact, we see only sterile workers which do not participate to the propagation of genes. The transmission is insured only by sexuals, queens and drones, which are far less numerous. Among them, a large proportion do not reproduce and the population size will depend only on sexuals which will transmit their genes to sexuals of the next generation.

A first approximation may be based on the number of colonies which are monogynous* and polyandrous*. The most cited number of drones which has inseminated a queen is 17, however the variation is very large for a few to about 30. In a population from south France (department of Lozère), the average number of patrilines* in 196 colonies was estimated to 13,8 (fig. 2). For various reasons, this number is slightly underestimated and we will retain 15 drones. The queen has two copies of each gene and the drones only one. The number of gene copies per colony is then 17. However, if each female of the progeny receives one or the other copy of the queen (probability 1/2 for each copy), it receives a copy of a particular male with an average probability of 1/15. The males receive a copy only from the queen. Consequently, the probability of paternity of a drone for a queen of the next generation is very low.

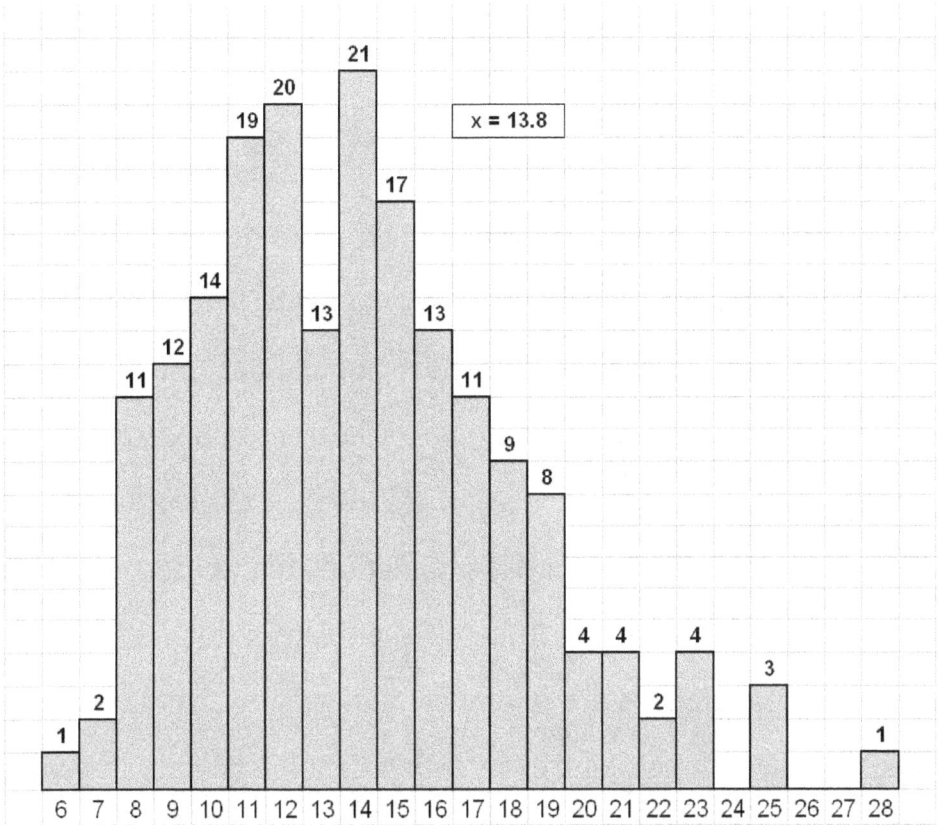

Figure 2 : **Distribution of the number of patrilines in 196 colonies from Lozère**

A calculation showed that the effective size (N_e) of a population is $1/N_e = 4/9 \, N_f + 2/9N_m$ where N_f is the number of queens and N_m the average number of drones which inseminated it [5]. From this formula, one can deduce the size of a population comprising for example 100 colonies (100 queens and 1,500 drones) : this is only 218. This value is calculated as if the two sexes where diploid for easier comparison with other species. It is roughly two times the number of colonies. If the honeybee was monandrous, the corresponding value should be 150. It is clear that the extreme polyandry of the honeybee do not much increase the population size.

In the preceding example, we have considered a population of 100 colonies, but the main problem is now to define a population, a unit of interbreeding individuals. The honeybees of a subspecies or from a country comprises numerous populations and, on the other side, the bees of an apiary belong to the same population than the neighbouring apiaries. Between these two extremes, the limits are difficult to delineate, except for populations living in isolated islands or valleys. In the general case, the distribution is continuous, neighbour populations are in contact and they exchange sexuals and genes. In these conditions, even with a detailed genetic or ecological study, the extension of a population is very difficult to appreciate.

However a particularity of the honeybee exists which can help us. The young, virgin queens are inseminated in particular places called "drone congregation areas" where the males recruited in a given perimeter converge. In these congregations, the males, whatever their origin, have all the same opportunity to inseminate a queen. This is one of the most perfect systems in animals and plants for an equal opportunity offered for reproduction because in almost all other cases, there is generally more chance to form the couples between individuals from the same limited area. The name for this effectively interbreeding community is a *deme*. Males and females may both fly over several kilometres and a perimeter of 10 km of diameter is about the surface of recruitment of these demes.

A genetic analysis was performed some years ago to estimate the number of colonies which delegate drones in a particular congregation area [1]. In the region of Frankfurt (Germany), at Oberursel, a pylon was built to analyse movements of queens and drones and sex behaviour. Several hundreds of males were captured at the tip of the pylon but only a fraction (142) was studied. 20 highly variable microsatellite markers were used. To estimate the number of colonies from which the drones came, it was necessary to classify them as brothers (from the same colony) and non-brothers (from different colonies). The rationale is as follows: if a male has a given variant for one microsatellite, a brother of this drone will have inherited the same copy from the same mother with a probability ½. In fact, the total probability is a bit higher because the second copy of the mother might be also the same variant if the female was homozygous for this variant. If the two drones are not brothers, the second one may have the same variant because several other mothers bear it. The probability that a male taken at random receives it is the frequency p of this variant in the population. This value is generally far lower than ½. With a single marker, one cannot distinguish brothers : even if the probability of being brothers is three times higher than non-brothers, the number of males carrying the same variant is high and the two categories cannot be distinguished on this base. However, the use of numerous markers approaches the solution because for brothers there is an accumulation of concordant markers which is not present in non-brothers.

In the samples we found 20 groups of two brothers, six of three and one of four, the remaining 80 were singletons (they had no brother in the sample), i.e. a total of 107 colonies. However, because only a small sample of the congregation was studied, numerous colonies were not represented by a drone. The distribution of the number of individuals per colony allows the estimation of this class and finally, the total number of colonies which had delegated drones to the congregation was found equal to 238. This value is rather close to the number of colonies living in the estimated perimeter of the congregation. The effective population size is a little more than two times the number of colonies and hence, close to 500. This figure may vary locally, depending on the landscape, the density of colonies, etc. It has to be noticed that the congregation delineates a deme rather that a population and the latter are more numerous but the exact size remains unknown; a figure of 1,000 is often retained without solid basis.

4.1 Factors modifying the effective population size
OUR HERITAGE

In spite of an intense movement of queens and colonies by man, the picture offered by the distribution of honeybees in Europe,

Middle East and Africa is rather close to the natural distribution. Due to recent climatic episode in Europe, the polymorphism of the populations is rather low, for instance in mellifera and ligustica. These two subspecies have expanded their territory recently from their refuge in South Spain and Italy, perhaps from small populations and the polymorphism was not restored during the last millenniums.

INCREASE

The development of a very dense apiculture could seem a method to maintain or increase the polymorphism. In fact, as stated above, following a population expansion, several thousands of generations are necessary to adjust the polymorphism to the population size. The positive point is that drift is reduced and hence, if there is no increase of the polymorphism, at least the decrease is stopped. Another positive point is that the changes of frequencies of alleles attributable to genetic drift at every generation are lower and in large populations a « purifying » selection may play its role in preventing the fixation of deleterious alleles.

Transhumance within a geographic region allows size increase of the population as long as the hive movements coincide with the date of reproduction. This may be beneficial because the populations have the same genetic profile but, exchanging genes, they may restore the presence of genes lost in the other populations. A negative effect may be the modification of local ecotypes*.

More problematic is the queen trading across countries or even continents. Generally, imported queens belong to another subspecies living in a very different environment. In that case, it has been generally observed that the proportion of foreign markers in populations is not very high and often far lower than the frequency expected from the rate of reiterated importation. For instance, after a long period of noticeable importation of *ligustica* in France, the proportion of mitochondrial DNA is generally limited to a low percentage. The *ligustica* signature in the nucleus is about of the same order. A possibility is that the imported genes have a low success, some favourable alleles might be introgressed but, for the moment, we are almost totally ignorant on these points.

In several regions, beekeepers have decided to create conservatories for the preservation of local honeybees. In many cases, given the low percentage of introgressed foreign alleles in the populations, this might be possible for almost all countries. In other words, these projects are not utopian because it is still possible to preserve local bees with a reasonable purity. However, the recommendations are very difficult in this way because these projects are always somewhere between romanticism and rationality. After all, the natural evolution of the species through millenniums was in fact punctuated by such events of massive introgression. For instance, in South Spain, the genetic profile of *iberica* is very close to that of *mellifera* for the nucleus but mitochondrial DNA is mainly

African. The same is true for *sicula* in Sicilia, which is nuclearly close to *ligustica* but also African for mitochondrial DNA. Continental Italia itself was during a rather recent episode of the ice ages colonised by an admixture of honeybees from central and Western Europe.

Returning to a pragmatic point of view, importation may lead to opposite results. Generally, when importation is low, hybridisation moderate and selection active, some adaptative genes of the imported race may introgress and this is beneficial for the local population. On the other side, with massive and reiterated importation, the gene flow may destroy the genetic architecture of the local race. An interesting situation is when virgin queens are imported (or produced locally) and allowed to be inseminated in the field. The hybrids are vigorous and the productivity of the hives headed by a queen of one subspecies (or a commercial line) inseminated by local males is excellent. However, with time, if hybrid queens reproduce, there is a risk to introduce the foreign genes within the local race. In that case, not only the efficiency of the process will decline with time but also the adaptation of the local population will be destroyed. Note that in these cases, the genetic variability is higher than in native populations, a feature that shows that high polymorphism is not always synonymous of higher performances.

DECREASE

As stated above if the increase of population size is not accompanied by an increase of polymorphism, the decrease of the size is responsible for a loss of variability, at least if this decrease does not cease rapidly. The rare alleles are the first ones to disappear. Rare does not mean useless (for instance, in the division of work, some rare tasks are useful).

Another consequence of the small population sizes is the possibility to fix deleterious alleles. The most deleterious ones are generally rapidly eliminated. However, some slightly deleterious alleles may persist in a population. If the population is large, their frequency will decrease by the action of the selection. If the population is small, the frequency will change anarchically at each generation under the effect of drift (change in frequency are higher by drift than those expected by selection). And finally deleterious alleles may be fixed. A small population has a reduced polymorphism and accumulates deleterious alleles in the course of generations. Most of the small populations will hence disappear.

It has to be noticed that selection and queen rearing are two apicultural practices that may reduce the effective population size.

5. CONSERVATION OF THE VARIABILITY

5.1 Several reasons why it is important to preserve polymorphism

The most obvious one is the sex locus: it must be heterozygote to produce a female. This is one of the rare well-documented cases were the heterozygote has an advantage (homozygote diploid drones are killed). In that situation, it has been shown that the number of variants must bee large and they have all the same frequency.

In other cases, variability must be maintained in order to have a diversity of genotypes and phenotypes within a colony. The enormous rate of recombination during queen meiosis (more than 100 chiasmata) is a source of diversity for the progeny. Similarly, the extreme polyandry* creates numerous genotypes, including the rarest. However, recombination and polyandry are unable to produce variability if there is not a diversity at the gene level in the population. The intracolonial genetic diversity is a very significant part of the diversity present in a population. It allows every colony to have a variability sufficient for aggressions (parasites, infections...) and ensure the diversity of tasks.

When the number of reproducers decreases (in the population or during a selection experiment for instance), the members of the couples become more and more related to one another and consanguinity increases in the progeny. Homozygotes for deleterious alleles appear and they are not protected by dominance. It has been repeatedly shown that the honeybee for numerous characters is very sensitive to consanguinity. One way to prevent consanguinity is a high population size.

References

[1]   Baudry E., Solignac M., Garnery L., Gries M. Cournuet J.-M. & Koeniger N. 1998. Relatedness among honeybees (*Apis mellifera*) of a drone congregation. *Proc. R. Soc. Lond.* B 265: 2009-2014.

[2]   Beye M., Hasselman M., Fondrk M.K., Page R.E., Omholt S.W., 2003. The gene csd is the primary signal for sexual development in the honeybee and encodes an SR-type protein. Cell 114: 419-29.

[3]   Bienefeld K. and Pirchner F., 1990. Heritabilities for several colony traits in the honeybee (Apis mellifera carnica). Apidologie 21:175-183

[4]   Chevalet C. and Cornuet J.M., 1982. Etude théorique sur la sélection du caractère "Production de miel" chez l'abeille. I. Modèle génétique et statistique. Apidologie 13: 39-65

[5]   Chevalet C. and Cornuet J.M., 1982. Evolution de la consanguinité dans une population d'abeilles. Apidologie 13: 157-168

[6]   Cornuet J.M. and Chevalet C., 1987. Etude théorique sur la sélection du caractère "Production de miel" chez l'abeille. II. Plan de sélection combinée de reines en fécondation naturelle. Apidologie 18: 253-266

[7]   Falconer D.S. and Mackay T.F.C., 1996. *Introduction to Quantitative Genetics (fourth edition)*. Addison Wesley Longman Ltd. 464p.

[8]   Hasselman M. and Beye M., 2004. Signature of selection among sex-determining alleles of the honey bee. P.N.A.S. 101: 4888-4893

[9]   Hunt G.J., Collins A.M., Rivera R., Page R.E. Jr, Guzman-Novoa E., 1999. Quantitative trait loci influencing honeybee alarm pheromone levels. J. Hered. 90: 585-589

[10]   Page R.E. Jr, Fondrk M.K., Hunt G.J., Guzman-Novoa E., Humphries M.A., Nguyen K., Greene A.S., 2000. Genetic dissection of honeybee (Apis mellifera L.) foraging behavior. J. Hered. 91: 474-479

[11]   Page R.E. and Laidlaw H.H., 1982. Closed population breeding: 1. Population genetics of sex determination. J. apic. Res. 21: 30-37

[12]   Solignac M., Vautrin D., Baudry E., Mougel F., A. Loiseau, Cornuet J.M., 2004. A microsatellite-based genetic linkage map of the honeybee, *Apis mellifera* L. Genetics 167: 253-262

[13] Woyke J., 1976. Population genetic studies on sex alleles in the honeybee using the example of Kangaroo Island bee sanctuary. J. apic. Res. 15: 105-123.

CHAPTER 3

REQUIREMENTS FOR LOCAL POPULATION CONSERVATION AND BREEDING

F. B. Kraus

1. APICULTURE AND BIODIVERSITY

Apiculture, the ancient craft of beekeeping, is practised by man since the early beginnings of civilisation. While still being a hunter-gatherer, man was also honey hunter collecting the honey and brood of wild honeybee colonies from trees and cliffs. Such honey hunting can still be found in some cultures in Africa and South East Asia nowadays. But during the course of civilisation most cultures developed, when they switched from hunter/gatherer to agriculture, more sophisticated techniques of beekeeping to ease the harvesting of honey. With the beginning of beekeeping also started the moving hives and colonies over large distances, since people took their bees with them when they set of to colonize new land. The most obvious result of this practise is the present worldwide distribution of the Western honeybee (*Apis mellifera*), which originally was restricted to Europe, the Middle East and Africa.

The honeybee is quite outstanding among the animals kept by man, since domesticated populations still share their gene pool* with feral populations. This is due to the complex mating biology of *A. mellifera*, where all drones and virgin queens of a population meet at so called *drone congregation areas* (DCA) * for mating, rendering it nearly impossible to keep managed colonies separated from wild colonies. Nevertheless, considerable time and logistic effort has been invested, especially in Central Europe, to achieve relatively "save" mating apiaries for bee breeding. Most of these mating apiaries were placed on isolated islands or in secluded mountain areas. Ironically the problem of keeping domesticated population separated from wild ones has recently reversed in many parts of Europe, where beekeepers are interested to conserve their local honeybee populations and prevent them from mixing with commercially imported breeding strains.

The requirements for both purposes, commercial breeding and conservation, are rather similar, since they have to take into account the same biological features and peculiarities of the honeybee, to achieve isolation from surrounding "unwanted" populations. The main differences between commercial bee breeding and conservation breeding is the amount of genetic diversity aimed for. Breeding of certain commercial bee strains requires focusing on a small part of the genetic diversity of a given population, because only a limited number of queens, which show the desired behavioural or colony traits will be used as starting population within a breeding scheme. This selection process unavoidably leads to a reduction of genetic diversity* (which is the essence of selection), resulting in a genetic and ecotype depletion also known from many other commercially breed organisms, e.g. cows and chicken, or rice and apples in plants. In contrast, the aim of conservation areas and breeding for conservation should be to maintain a genetically diverse population, thus involving as many colonies as possible to have large population sizes. The larger a population, the smaller the chances of loosing genetic diversity by stochastic processes, population bottlenecks* or diseases.

2. MATING RANGES AND INBREEDING

One of the most basic features in the mating biology of the honeybee is that both sexes, the drones as well as the virgin queens, leave their mother colonies to mate in midair. Since this behaviour posses a considerable risk to the virgin queens of not returning safely to the nest scientists were puzzled why the queen should not simply mate in the hive. The unusual mating behaviour was mainly explained by the need of the queens to avoid sister/brother matings. Sister/brother matings are extremely detrimental for honeybees because sister brother matings yield the danger of inbreeding, which results in the "shotgun brood" phenomenon. If a female honeybee carries two identical copies of the single sex determining gene, it will become a functional drone, which is recognised and cannibalised by the other workers during larval development. If such diploid drones occur at high frequencies in a colony, a reduced performance and survival rate of the colony as a whole is the consequence. High proportions of diploid* unviable drones arise from inbreeding, because the chances of having identical sex determining genes are enhanced among close relatives. The result is that every second egg is unviable resulting in the shot gun pattern of the sealed brood comb. In fact the mating system of the honeybee, with drone congregation areas (DCA), where all drones and queens from given population gather, result in nearly completely mixed, so called "panmictic*" population structure, with a very low risk of inbreeding. The more colonies participate in a given DCA the smaller the changes for a virgin queen of mating with a brother drone.

The distances which drones and queens are willing to cover, to reach a suitable DCA can be considerable. Virgin queens have been reported to fly up to eight kilometres if necessary, a distance even exceeded by the drones, which are known to fly up to ten kilometres to a DCA for mating. In extreme this can result in an mating area covering more than 300 square kilometres. Of course these immense mating ranges demonstrate the high potential of the honeybee to avoid inbreeding by minimising the changes of mating between closely related individuals, but they are also the main problem in honeybee breeding, where the aim is to exclude mating with and genetic introgression* from surrounding populations. It can easily be imagined that in more densely populated regions, with many beekeepers, it is virtually impossible to control the origin and composition of all present honeybee colonies. This can only be achieved when the vast majority of beekeepers in a given region agree in keeping and breeding bees of a particular race or ecotype. Even in rural or "wild" areas where managed colonies of unwanted origin might be excluded, wild colonies are still not controllable and are the main threat for commercial bee breeding. The latter might not be the case for conservation efforts, because local honeybees are more likely to survive in the wild without any attendance of a beekeeper than escaped swarms from commercial beekeeping (but this strongly depends on the races used).

A further feature which complicates the isolation of populations is the need for many commercial beekeepers to move hives on a massive scale during seasonal migrations. Migratory beekeeping drastically enhances the geneflow even between distant populations, when drones of such mobile colonies are not prevented from leaving their colonies and mate with queens from the visited areas. How can we reconcile the economical need for commercial beekeeper to visit rich honey flows with the need of conservationist beekeepers to maintain local endemic races?

2.1 Ways to achieve isolation

There are a series of proven apicultural routines to keep gene flow at a minimum level and to achieve reasonable isolation from surrounding populations. Since the problem of isolation applies to whatever breeding program desired (conservation or commercial beekeeping), there have been many, quite successful, attempts to establish isolated mating apiaries in commercial beekeeping during the last century. The best examples for such mating apiaries can definitely be found in Germany, where breeding programs for the *Carniolian* bee, *A. m. carnica*, have been launched in the first half of the 20th century and were maintained and improved until today (the whole German breeding program was "successful" resulting in the extinction of *A. .m. mellifera*). Traditionally islands and secluded mountain areas have been favoured and used as mating apiaries to achieve isolation from the surrounding native *A. m. mellifera* populations, for the breeding and improvement of certain *Carniolian* breeding lines. The basic assumption here is, that queens and drones are likely to avoid the crossing of geographical barriers like mountain ridges and longer stretches of open water. But also so called "pure breed areas" have been established e.g. in Bavaria, with no special geographic barrier to keep populations apart.

ISLAND MATING APIARIES

In Germany the mating apiaries on island have been established on North Sea islands along the coast. These islands are between 5 and 10 kilometres away from the mainland and are kept otherwise free of any other honeybee colonies, than the ones brought there for breeding. It is immediately obvious that, even if we assume pretty good isolation, such island mating apiaries require a huge logistic effort to be established and maintained. Even for the honeybee colonies themselves, North Sea islands are quite unfavourable environments and they can only be kept there when they are properly cared for. These disadvantages might be avoidable if conservation areas for honeybees are established on larger islands or on islands with a more suitable climate, a sufficient supply of flowering plants and a prolonged flowering season.

Figure 1 : **Mating apiary on the North Sea island Spiekeroog (photo by Dr. Ralph Büchler).**

MOUNTAIN MATING APIARIES

Mountain areas are another possibility using natural barriers to keep honeybee population apart from each other. First, mountain areas are normally much less densely populated then lowland areas, and as a consequence it is much less likely to get genetic influx from unwanted breeding strains or ecotypes. Honeybees avoid the crossing of mountain ridges and seem to prefer to follow along a valley, ending up in establishing very stable DCAs over many years, as we know it from Alpine regions. In fact stable, localized DCAs might be an artefact of mountain or hill landscapes and it is unclear whether clear, distinct DCAs arise in geographically "profile less" lowland areas. Compared to island mating apiaries, the advantage of mountain areas is that the logistic effort is much less intensive. Some extreme locations, however can only be accessed by cable car or on foot.

Figure 2 : **View from the Salzkopf mating apiary in the Hunsrück mountains, Germany (photo by Dr. Ralph Büchler).**

PURE BREED MATING REGIONS

The concept of "pure breed regions" is probably the logistically most sophisticated approach to achieve mating isolation. Here the idea is to control the population by providing every beekeeper in at least 20km diameter with queens of a given breeding strain to assure that only drones and virgin queens of that particular strain do participate in the DCAs within the pure breed area. It may be difficult to realize such an approach for a conservation breeding program, but clearly it is not impossible. It has been successfully implemented in some Bavarian regions where beekeepers were particularly well organized and received assistance from governmental institutions. So far this concept was practicable because beekeepers in these areas were given "high quality" breeding strain queens for free, when they cooperated, and additionally these pure breed areas were enforced by legislation (for the beekeepers which do not agree in such breeding concepts voluntarily). The first stimulus, giving free queens to beekeepers, relies on the ability to convince all beekeepers in a given area that breeding or conserving a particular ecotype is a good idea, and it might be hard to achieve that. Additionally legislative regulations can implement mating regions. However, this is also difficult since it will always lead to increased and rather unpleasant conflicts between beekeepers (as already happened on the Danish island Læsø) and the enforcement of such legislation is difficult if not impossible to obtain.

3. POPULATION SIZES AND MATING FREQUENCIES

A crucial requirement for sustainable breeding of native honeybees is a sufficiently large population size. The smaller a given population the higher are the chances of inbreeding and the random loss of genetic diversity. Like in all social insects, it is not the number of individuals which characterises the size of a population. It is the number of queens (which equals the number of colonies in case of the honeybee) and the number of drones they are able to mate with. This effect is further enhanced by the haplo/diploid population structure* of honeybees, which leads to a decrease in the genetic, so called "effective" population size (for a more detailed description of the theoretical background of effective population sizes see chapter 2 of this book). The complimentary sex determining mechanism of honeybees with its selection towards a high rate of heterozygotes* within a population further reduces the effective population size.

The question of how many colonies are necessary to sustain a viable population of honeybees is difficult to answer and is highly dependent on the local conditions. It is not only the number of colonies but also their evolutionary and genealogic history which is important, the more genetically diverse a population is in the start of a breeding program, the smaller it can be in terms of the number of colonies. Examples where populations seem to be stable for several decades are found on the Danish island Læsø and the Australian Kangaroo island. But in both cases one has also to take into account that we do not know how strong an influx of honeybee colonies from outside the islands has been during the history of these populations.

A practical approach of assuring an optimal possible effective population size with a given number of colonies is trying to get a sufficiently high average number of matings per queen. The effective population size rises with the average number of matings achieved by queens in a given population. Honeybee queen generally mate with a high number of males on their mating flights, but the number of matings can vary considerably between populations depending on local environmental conditions. Thus providing good mating conditions for queens is a way to assure optimal population sizes. A visualisation of this effect is given in fig. 3. As it can easily be seen from the graph the effective population size rises quickly until it reaches a point of minimal returns. With about 10 matings per queen 95% of the maximum possible effective population size with an infinite number of matings per queen is reached . Thus by assuring that average mating frequencies are above 10 matings, one can make certain that population sizes are not reduced by this effect and prevent loosing genetic diversity due to an insufficiently high number of matings by the queens.

The question of how many colonies are necessary to start a conservation program with is hard to predict, since the answer is highly dependent on the history of a given population. Drastic declines in population size in historical time may already have lead to a severe reduction in genetic variability in a given population. From experiences with breeding of other endangered animal species (mostly mammals) we know that whole species can recover from amazingly few animals, but empirical examples are rare and in case of the honeybee the sex determining mechanism is definitely a severe restriction on to small populations. Since the number of sex determining alleles is the ultimate limiting factor for the viability of a given population, estimates of the number of sex alleles present in a starting population might be useful. But studies on this particular subject are far from being conclusive so far and rapid methods to evaluate this via molecular DNA methods are not developed yet. So the answer to this question might simply be: "the more the better", and hoping for a good outcome.

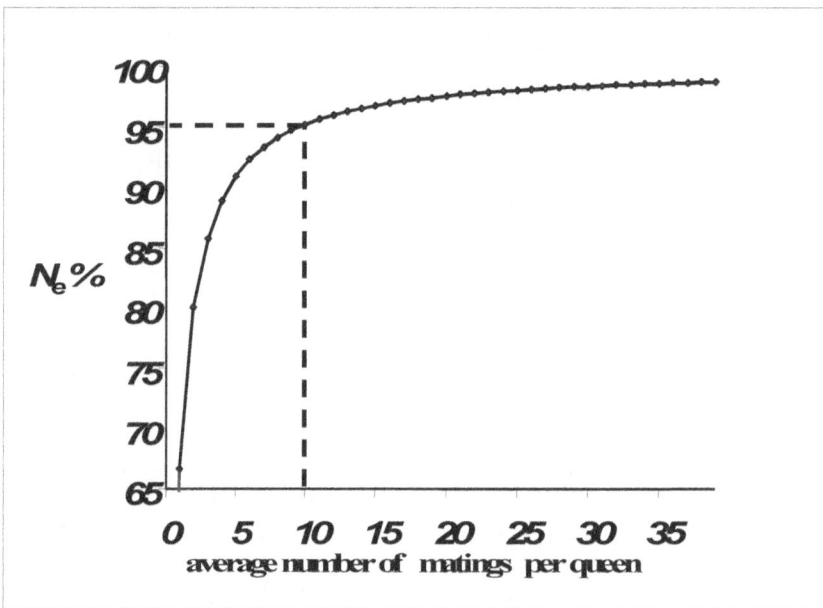

Figure 3 : **The relationship between the average number of matings of queens in a population and the effective populations size (N_e). With an average number of 10 matings 95% of the maximum possible effective population size are reached (doted line).**

3.1 Impact of environmental conditions on mating

It is well known that the main environmental factors influencing the mating success and frequency of honeybee queen are the weather conditions. For their mating flights queens have only a limited time window of a few weeks after emergence. In case they have not been able to mate during that time window, they will start laying drone brood and will be eventually replaced by the workers. Besides the fact that queens are more likely to get lost under bad weather conditions, they also achieve much less matings under unfavourable conditions than queens under favourable conditions. For some North Sea islands mating frequencies as low as three matings per queen have been recorded using molecular paternity testing methods. Queens mating under favourable conditions on the mainland during summer achieve easily up to 30 or more matings. Data on matings in mountain areas obtained so far indicate that here reasonable mating conditions can be achieved with mating frequencies between 10 and 15 matings per queen. Not only differences in weather conditions, but also behavioural differences between subspecies can result in different average mating frequencies. A graphic illustration of different mating frequencies under different environmental conditions in Germany is given in fig. 4.

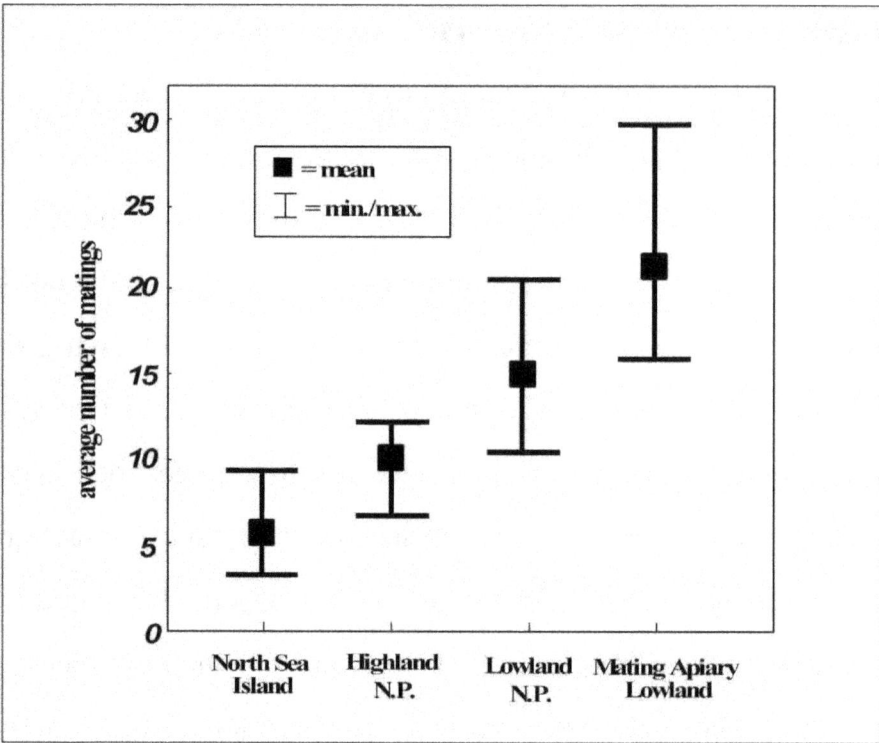

Figure 4 : **Mating frequencies in four different populations in Germany. Given are the average (mean) mating frequencies and the minimum (min.) and maximum (max.) number of matings found in the given population. North Sea Island = Island of Neuwerk; Mountain N. P.= Hochharz National Park; Lowland N. P. = Müritz National Park; Lowland Mating Apiary= Schwarzenau/Bavaria.**

The crucial question for the establishment of a conservation or breeding area of course is, which are the weather parameters with a strong impact on the mating frequencies? Analysis of weather parameters involved, revealed that high wind velocities seem to be the major problem for the queens during matings, as long as the temperatures are high enough for the queens to leave the hive and the mating season does not take place during a two week non-stop raining period. For other weather parameters (rainfall; humidity; temperature) no significant impact on mating frequencies was found and they seem to play only a minor role. Figure 5 gives a graphic illustration on the negative correlation between wind velocities and the mating frequency.

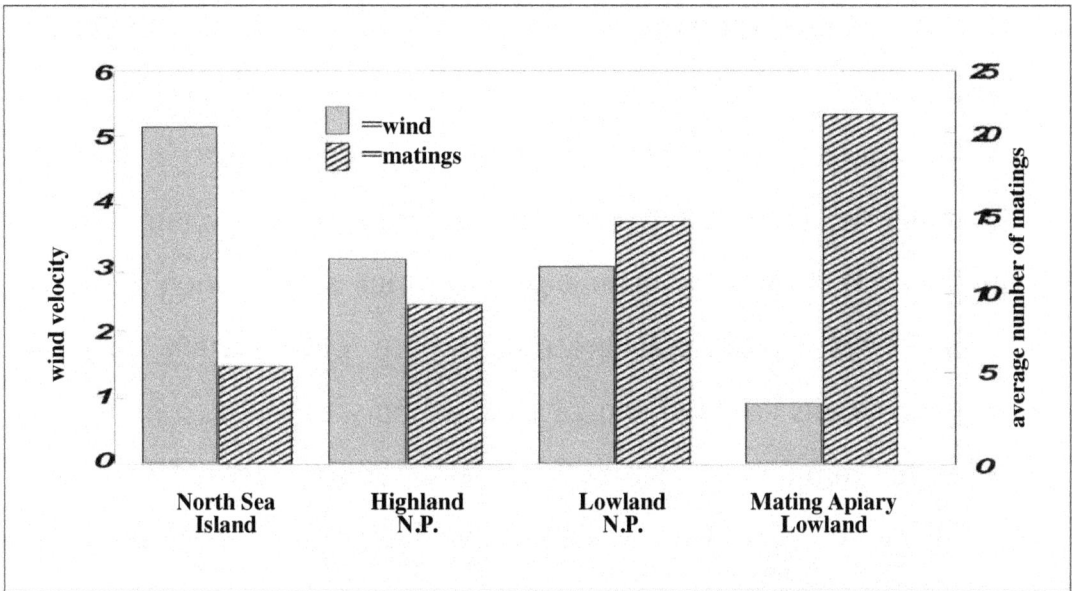

Figure 5 : **Average mating frequencies and average wind velocities during the mating season in the four studied populations in Germany. Wind velocities are given in meters per second (left x- axis), the average mating frequencies are given on the left x-axis. North Sea Island = Island of Neuwerk; Mountain N. P.= Hochharz National Park; Lowland N. P. = Müritz National Park; Lowland Mating Apiary= Schwarzenau/Bavaria.**

Clearly the impact of wind is the major problem for most island or coastal areas, which are regularly affected by strong gales, but again size of the island or the presence of wind breaking structures like forests might be helpful in preventing a too harsh impact of wind velocity on the matings. Another side effect of very low mating frequencies is that queens can run out of sperm supply after one or two seasons, an effect which sometimes seem to occur with commercially island mated queens.

4. NUMBER OF DRONE COLONIES PARTICIPATING IN MATING

The more colonies of a given population participate in mating, the better the chances for preserving a high genetic diversity are on the long run. This is the main aspect in which conservation breeding differs from commercial breeding where only certain selected breeding strains are bred and genetic diversity is kept at a minimum level. In commercial mating apiaries therefore only few drone producing colonies of a given breeding line are used to produce the drone which are to mate with the virgin queens (e.g. only eight colonies on some islands). During the first trials to establish island mating apiaries in Germany only a *single* drone producing colony was used for supplying the mating apiary and the corresponding DCA with drones. As a consequence the results achieved were rather poor, with many queens failing to mate at all, getting lost or starting laying drone brood after a short time. After it was realized that the reason for this problems was the insufficient drone supply the number of drone colonies was drastically increased resulting in better mating success of the queens.

Genetic studies of the mating success of drone colonies on mating apiaries have revealed that drone colonies do not contribute equally to the matings at a given mating apiary. Since some of these colonies, even if they are kept under equal conditions fail to successfully produce drones, the *effective* number of drone colonies is much smaller than the actual number of drone colonies present. As a rule of thumb, one can calculate that the effective number of drone colonies is only half the actual number of colonies at a mating apiary or in a population. Figure 6 shows the contribution of eight equally strong drone colonies to the matings on the Island of Neuwerk, a commercially used mating apiary, during a matings seasons. Under natural conditions or without beekeeping measurements we would expect to find even larger differences between drone production and male mating success of colonies. This is phenomena should be kept in mind if establishing any protection area or mating apiaries for breeding of native honeybees.

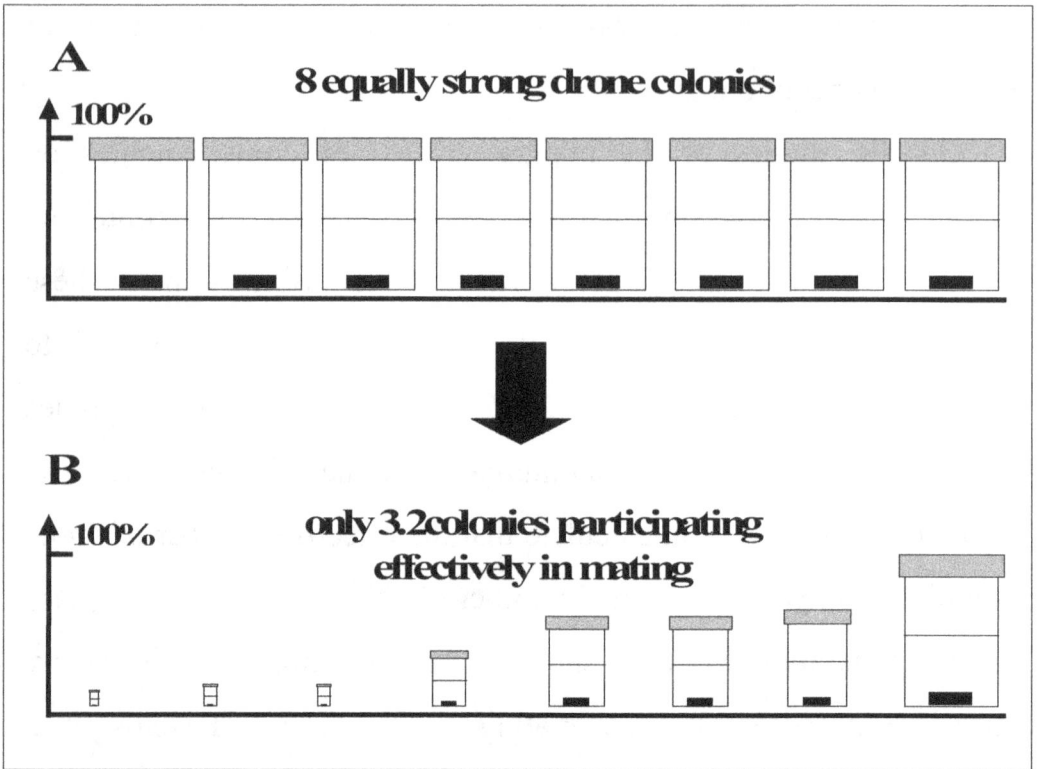

Figure 6 : **Eight equally strong drone colonies (A), which were put on the island during mating to provide an ample drone supply. The participation in mating of these colonies (B) is however far from being equal among the colonies, and some nearly fail to reproduce at all. Effectively only 3.2 colonies (not the numerical eight!) participate in mating on the island.**

5. RECENT LESSONS FROM COMMERCIAL MATING APIARIES

Since islands have long stretches of open water between them and the mainland populations, they were considered so far safe in terms of mating isolation. Recent DNA fingerprinting studies indicate that under extreme conditions (where no drones are available) queens also cross larger stretches of water. Studies in which foreign matings were detected on islands are very likely to have resulted from management problems, rather than from influx of drones from the mainland.

DNA studies from a Mediterranean island (Unije), which was used for *Varroa* mite resistance breeding, indicate that the whole island of Unije contains a single panmictic population, in spite of some geographical barriers present. Since this island roughly covers an area of 17 km^2 this again shows the high potential of honeybees to cover large distances fro mating.

6. CONCLUSION: CONSEQUENCES FOR HONEYBEE CONSERVATION AREAS

Taking into account all the experiences gained so far from commercial bee breeding and the peculiar mating and populations biology of the honey bee, we can conclude that 100% isolation from surrounding populations is very hard to achieve. However, gene flow between populations is a very natural phenomenon and as long as it can be kept at a reasonable level it will be acceptable for conservation breeding and the implementation of protection areas for native honeybee populations. Within a healthy and prosperous population with strong drone producing colonies some foreign colonies sending drones to the DCAs should simply be outnumbered. Thus the case of conserving native honeybees is definitely not hopeless, it mostly will depend on the political will of the majority of beekeepers in a given area to cooperate in the conservation efforts.

As we saw in this chapter, islands are the most easy to control and probably the best isolated areas one could use for the establishment of conservation areas. But the major setback of islands is that they are often quite small and therefore not capable of supporting large and viable populations of honeybees, which are less vulnerable to the effects of random loss of genetic diversity. A further factor, at least in colder climates, is that islands are usually much more exposed to unfavourable weather conditions than more protected mainland areas, thus leading in turn to unfavourable mating conditions. Also the logistic effort for keeping honeybees on islands can be much larger than for mainland areas of comparable size as long as the beekeepers do not live on the islands, like they do on the island of Læsø in Denmark or Kangaroo Island in Australia.

If the aim is to *breed* and produce queens of a given endangered population or subspecies, the size of the islands is not so important any more, since a population viable on the long run is not required. The only requirement besides good mating conditions would then be a high number of drone producing colonies during mating season to ensure a starting gene pool diverse enough to ensure a high genetic diversity.

Mainland areas are clearly preferable for the establishment of conservation areas from available range, mating conditions and infra structure. Here the major setback is the problem of isolation. In countries thinly populated, where it is possible to assure beekeeping free regions in at least a 40km diameter, the mainland is clearly preferable. It has to be kept in mind that in this case influx from bordering populations is always possible and colony numbers should be as high as possible, since this will lead to an dilution effect which reduces the impact of geneflow.

In case of more densely populated mainland areas, one might still try to establish conservation areas in secluded mountainous areas. Mountain regions which are usually less populated even in highly industrialised countries in central Europe. Such regions should provide weather conditions good enough for a sufficiently high average mating frequency in combination with reasonable isolation. It is also possible to keep a large number of colonies to ensure a high genetic diversity in the population on the long run. Sometimes the human population in remote mountain areas might also be more willing to conserve traditional ways of beekeeping or native bee races than beekeepers in more "modern" parts of a given country, since rural areas are often known fro their general more "conservative attitude".

CHAPTER 4

PRACTICAL ASPECTS OF BEE BREEDING FOR BIODIVERSITY AIMS

Marco Lodesani, Cecilia Costa

1. BREEDING QUEENS AND DRONES AND PRACTICAL CONSERVATION TECNIQUES

The assumption underlying this chapter is that an ecotype has been determined and described and that it is recognisable by biometric, genetic and behavioural characters.

As in any kind of genetic improvement programme the first step of a conservation programme is evaluation of the individuals. To ensure the conservation programme a chance of success it is important that economic characteristics be evaluated on the same level as the racial standards. Also the peculiar behavioural characterstics (e.g. overwintering) must be considered, especially in function of adaptation to the local climatic / geographic / parassitological conditions.

The colonies which correspond to the racial standards and to the chosen commercial and ecological characteristics are selected for queen and drone rearing. Differently from traditional breeding programmes the number of colonies selected for reproduction must not be limited to only a few individuals: in this way we would risk losing the genetic variability inherent to the population which is object of the conservation programme, with detrimental effects to its survival in the long term (for details on the importance of genetic variability and population size see Chapter 2).

1.1 Queen rearing

The first point to consider must be location of the breeding yard: it must be exposed to the sunshine and protected from dominant winds, and it should be able to provide constant nectar and pollen sources throughout the breeding season (to the benefit of the breeders pocket, who would otherwise spend much time and money on feeding, and of the quality of the produced queens). Due to the intensive modern agricultural techniques and the use of crops with low nectar production, planting nectar and pollen producing species in the area surrounding the breeding station can be a valid aid to the breeding and a good investment.

According to natural rythms queen rearing begins when drones start emerging, in this way queens will be born and ready for mating when the drones are sexually mature. For a breeder it may be important to anticipate or to postpone the production of queens, in which case the breeding of drones must be accordingly planned (fig.1).

Insert | Queen
drone | lays
comb | eggs

Transfer of sealed
drone brood
(if necessary)

Drones
emerge

Insert drone | Insert special
cage at hive | drone flight
entrance (II) | super (II)

0   1   12   23   25   28   34   39
(x-39)   (x-26)   (x-15) (x-13)   (x-10)   (x-4)   (Day x)

SEXUAL
MATURITY

Transfer
cells into
nuclei

Queens
emerge

Treatment
with
CO_2 (II)

Grafting

0   10   11-12   17   18
(x-18)   (x-8)   (Day x)

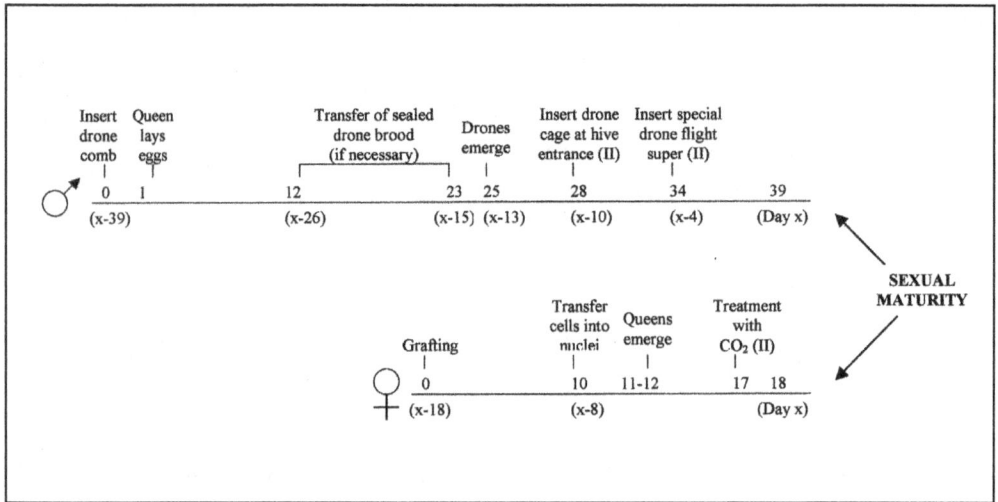

Figure 1 : **Planning calendar for sinchronised production of sexually mature drones and queens.**

The basic material necessary for queen rearing is:

- *Queen cell cups*, in which the larvae are grafted. They are a reproduction of the initial queen cells in the hive. They can be made of wax or plastic, are 9mm wide and 8mm deep. The wax cell cups can be home made by dipping a forming stick into melted wax and letting the wax solidify around it.

- *Queen cell bars*, to which the cell cups (10-20) are attached. They are long enough to be fitted between the end bars of a standard frame. The cell cups can be directly attached to the bar with melted wax or with some kind of medium to facilitate removal of cells, such as metal strips fixed to the bar or specially made plastic stands;

- *Queen cell frames*, in which the cell bars (1-3) are inserted. It is the same size as a standard frame, with special gaps along the sides to fit up to 3 cell bars and a feeder at the top;

- *Queen cages*, to collect, transport and insert the queen into the hive. There are hundreds of different kinds which all have in common enough space to host the queen and 6-10 worker bees, an exit hole adjacent to a candy compartment and at least one surface with holes, to provide ventilation and protected contact during introduction. The worker bees have an important role in climatising the atmosphere inside the cage rather than of feeding the queen (in captivity the queen feeds herself). The cages can be built in wood or plastic (special kinds of plastic which do not create electrostatic energy which could damage the queen);

- *Candy*, it is used in the queen cage to provide the necessary nutrition for the bees and is useful when introducing the queen into the hive: the time it takes for the bees to eat the candy and free the exit hole are important in increasing probabilities of queen acceptance, especially if the worker bees are not removed. The best candy for queen cages is obtained by mixing honey (~15%) with powder sugar (~85%);

- *Grafting needles*, the instruments used to transfer the small worker bee larva from the frame cell to the cup exist in various shapes and sizes. In the metal model (similar to a dentist's instrument) the tip of the hook is rounded so that it does not dig into the cell bottom when slipped under the larva. The chinese models have a soft and flexible tip made of goose cartilage, that diminish risk of damagement of the larva, which is picked up together with a bed of royal jelly; they also have a spring system to facilitate deposition inside the cup;

- *Selected queen mothers*, from which to collect the larvae (production of which can be stimulated by feeding) and queen-rearing colonies.

GRAFTING

The optimal conditions for the grafting procedure are about 50% humidity and 26°C to prevent dehydration of larvae. Light is best obtained with a fluorescent lamp, which does not emit heat that might injure the larvae.

According to the kind of needle used the cups are primed with royal jelly (at room temperature, pure or diluted) to aid deposition of the larva from the picking instrument and to prevent dehydration (against which it is best to cover the bars containing the grafted cups with a wet cloth).

The principle by which the rearing of queens from worker bee larvae is possible is the effect of the different nutrition on larval development (a queen larva is fed with royal jelly throughout her development, whereas after the first 3 days the worker larva is fed with pollen and honey). A worker larva inserted into a queen cup and fed with royal jelly during its whole development cycle will become a queen. The earlier the process starts, the higher the quality of the queen: after the first 24 hours a queen larva is fed with higher amounts of royal jelly compared to a worker larva. Therefore to obtain good queens and to facilitate working procedures (simoultaneous emergence of the queens) it is advantageous to graft larvae not older than 24-36 hours. At this stage they are very small and almost transparent but visible with good lighting and a bit of practice (fig 2,3,4). The grafted larvae are fed by nurse bees in queenless or queenright colonies, especially prepared. In some cases the whole

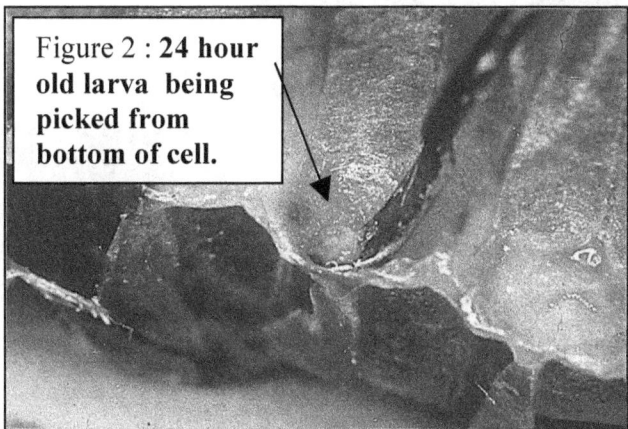

Figure 2 : **24 hour old larva being picked from bottom of cell.**

development of the queens is carried out by a single colony, in others several kinds of breeding colonies are used for the different stages of development ("starters" for acceptance and building of the cells and "finishers" for feeding them until they are sealed. The choice of the kind of breeding colony depends on race of bee (some races will accept queen cells more easily than others), on the climatic conditions (acceptance of cells is more difficult at the beginning and end of season) and on the beekeeper's organisation and personal preference. There isn't a better kind of breeding colony than another: as long as the colony is well fed and well populated, especially by nurse bees, the acceptance and rearing will be successful. Here a few kinds of breeding colonies are described, classified as "starter" or "finisher" colonies, but they can both be used for the whole process [12].

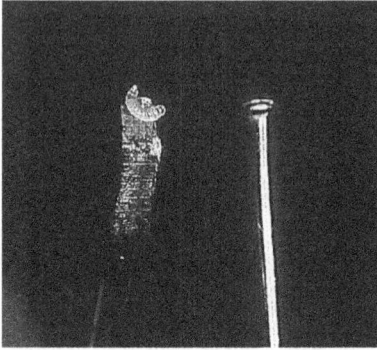

Figure 3 : **Picked larva (compare size of larva with size of pin).**

Figure 4 : **Larvae grafted into the queen cell cups on a drop of water and royal jelly.**

CELL-STARTER COLONIES

Specialised colonies used to feed the grafted larvae during the first 24-36 hours. They are made up of a large amount of young bees for which contact with their queen is interrupted, either by dequeening (queenless breeder colony) or by creating a breeding chamber separated from the queen with a queen excluder (qeenright breeder colony). There are many different ways of setting up breeding colonies which vary according to the kind of equipment used, to the race of bee, to the number of queens to be produced, but for all systems the important things are that the breeding colonies be strong and constantly provided with honey and pollen, so as to have many nurse bees well able to secrete royal jelly from their hypopharingeal glands.

- Queenless colony

It can be obtained by dequeening a colony, subsequently brood must periodically be added to ensure the constant presence of young bees. Either sealed or open brood can be inserted next to the grafted cells, the latter provides the advantage of attracting nurse bees but the hive must then be checked for "wild queen cells".

- Queenright colony

A brood chamber separated by a division board with a small piece of queen excluder is set up either above or next to a strong colony. Part of the bees of the colony will feel queenless and will be inclined to rear queens. The number of queens cells that can be obtained with this method are inferior compared to the previous method (~15 against ~30) but it has the advantage of being self-sufficient (the brood to be added in the queenless part is transferred from the queenright part) and of working like a normal honey gathering colony.

FINISHERS

Specialised queenright colonies used to feed the "started" larvae during the last 60-72 hours before the cells are sealed. These are normally used to free the space in the starter colonies for further grafted larvae to be accepted. The finisher colonies would accept a lower number of larvae compared to the starter colonies, but once the cells have been started the finisher colonies will continue feeding all the larvae. Here are some kinds of finishers:

- 10 frame finisher hive

A standard 10 frame hive can be used by creating 2 compartments with a vertical queen excluder and with modified independent roofs for either section. When the colony is well developed it can be divided with the excluder, placing 2 frames of unsealed brood and a feeder frame (filled with stimulant syrup) in the queenless area. After 24-36 hours the queen frame containing 2-3 queen bars replaces the feeder frame. The bars can contain grafted cells or accepted cells from a starter. Every 2 weeks the queen is placed in the orphan area to provide the young brood necessary to attract the nurse bees. This operation can be timed so that it takes place between a removal of mature cells and introduction of new cells.

- Vertical finisher

Two empty supers provided with an entrance hole or a standard bottomless hive are placed on top of a strong colony, separated by a queen excluder. Two frames of honey and pollen and 2 frames of emerging brood are transferred from the nest to the supers, placed on a side and provided with stimulating feed (in a feeder frame). If the nights are cold it is advisable to place a plastic sheet over the part of queen excluder not covered by frames. After 2-3 days a bar with 15 started cells is inserted in the upper unit, which will then be ready to accept another bar (also for starting) after 3-4 days. Once a week a frame of emerging worker brood is transferred to the upper unit from the nest , in exchange for a frame of honey (useful as stimulant in the spring). According to the development of the colony foundations can be inserted in the upper unit, attracting bees from below; in this case renewal of emerging brood can be carried out every 2 weeks. Frames should be checked weekly for presence of natural queen cells. This kind of finisher is not as productive as others but is adaptable to various field conditions (it can easily be reduced or widened, a super can be inserted between the 2 units when there is a strong nectar flow, the lower and upper unit can be exchanged to provide extra bees) and is relatively simple to prepare.

- Super finisher

This system works when a nectar flow is in course and the super already full. A queen excluder must be inserted between nest and super, then a super frame is removed and in its place a bar frame is placed (if necessary, modified into super size) with the started queen cells.

When setting up or renewing any of these breeding units utmost care must be given to not inadvertently introduce a queen, for this will strongly diminish success. Attention must also be dedicated to timing of queen emergence, for one virgin queen can kill all other unborn queens in the breeding unit.

HANDLING OF QUEEN CELLS

The sealed queen cells can either be:

a) left in the original starter or in the finisher (suitable if the rearing cycles are distant in time)

b) transferred to a second finisher that acts as "cell bank" (2 frames of brood and 2 of honey and pollen will be sufficient to provide the suitable temperature and humidity for about 100 cells)

c) transferred into an incubator at 34-35°C and 60-70% relative humidity

If the cells are moved (b and c) this can be done without risk of damaging the queens between the 5th and 7th day from grafting. On the 8th or 9th day from grafting the queen detaches herself from the top of the cell: any movement during this stage can damage her. On the 10th or 11th day the cells can again be moved. When transferring the cells they must always be kept in their natural vertical position; they can be placed in polysterene containers with appropriately sized niches, covered by a damp cloth. Temperature is very important in queen development: temperatures below 34°C progressively increase development time (cells kept at 30°C from the 5th day from grafting will emerge 3 days later than expected) whereas a slightly higher temperature will anticipate emergence.

QUEEN-MATING NUCLEI

These are small queenless colonies in which the ripe queen-cell is placed one day before expected emergence. There are many different kinds of queen-mating colonies, suitable for different kinds of climatic conditions, organisation of the farm, beekeeper's preference. If the queens have been raised for substitution in the beekeeper's own hives the mating colony will be the queen's final destination; if the queens have been raised as part of a commercial programme they will then be removed from the mating colony after beginning of oviposition, and another ripe cell will be inserted in its place. According to the kind of use that is required from the mating colony (number of cycles) and the climatic conditions in which it is placed a beekeeper may choose among many different kinds queen-mating colonies. Whatever kind is chosen must be prepared in such a way as to have enough bees of different ages to provide a constant temperature of 30-35°C even in bad weather conditions and food for the nurse bees to provide the queen with the necessary nourishment. The number of bees of course depends on size of the mating box: in the smallest commercially available "baby nuclei" the minimum number of bees necessary to raise a queen is 500 but this number must increase to several thousands in bigger nuclei (fig.5).

Figure 5: **Polistyrene mini nuclei ("Kirchhainer" on the left, "Apidea" on the right).**

PREPARATION OF QUEEN-MATING NUCLEI

To facilitate formation of the mating colonies many beekeepers prefer those kinds of boxes in which standard hive frames can be inserted, such as standard hives, supers or swarm boxes with internal partitioning. Such colonies, being bigger, are also strong enough to overwinter and don't need to be fed as much as small ones.

When making up small colonies the combs can be either cut out from standard hive combs or built directly from the bees in the small nucleus. The bees necessary to fill the nuclei are shaken from regular colonies into a ventilated box, where they are fed with syrup and sprayed with water or syrup to facilitate distribution into the boxes. Alternatively the nuclei can be filled by shaking bees from regular colonies through a giant funnel (a comb covered with bees can be sufficient to populate 3 mini-nuclei) but this technique has the disadvantage of distributing the bees unequally in terms of age among the nuclei, as the older bees are the first to fall when shaken. Another technique used by some breeders to fill the nuclei with bees is anaesthetisation with carbon dioxide (CO_2). With this method,

120

compared to spraying with water or syrup, there is more time available for the beekeeper to pour the bees in the boxes; also the bees do not risk not being able to dry themselves if the weather turns bad. Once filled, the nuclei are left for 24-48 hours in a cool dark place before being taken to the mating station.

Baby nuclei are useful if a large number of queens is produced: not many hives are needed to form a large number of mini-colonies, they can be easily transported to and from the mating station, individuation of the queen is easier. To give satisfactory results however, they must be carefully managed. Any climatic change influences the small colony's balance, and the beekeeper must be quick in providing extra feed or in joining the smallest colonies when the weather conditions are bad, and quick in adding foundations when there is a strong nectar flow. The timing of removal of egg-laying queens and introduction of new cells must be carefully respected, as permanence of the queen in the mini-colonies should be no more than 2 weeks (to avoid the risk of swarming due to the restricted space).

Therefore, depending on the time and expertise available to manage the nuclei, the number of queens produced and the number of planned cycles, the kind of weather and the necessity of overwintering, the amount and intensity of nectar flows, the beekeeper shall choose the kind of queen-mating colony [4].

1.2 Drone rearing

In a conservation programme it is essential to ensure maintenance of genetic diversity: drone rearing must be stimulated and encouraged in as many colonies possible - according to the selected traits. The mating environment should be saturated with the selected drones and it is advisable to use drones produced by the natural mated daughters of the mothers chosen for drone breeding because these drones will faithfully reproduce the genetic characteristics of the "grandmother" queen [1a]. The drone colonies should be well fed at the end of winter so as to stimulate the worker bees in the construction of special drone foundation sheets or alternatively a previously built drone comb can be inserted next to the external brood. If the seasonal conditions are not adequate for drone egg laying (bees naturally tend to produce drones in spring), the queens can be confined to the drone comb by use of a cage made out of queen excluders (fig.6). After 2-3 days deposition should be more or less complete and the queens can be freed. When the drone brood is sealed (after ~10 days) it can be transferred to a different colony and the queens can again be caged for further drone oviposition (the presence of drone brood inhibits male oviposition). The colony in which the sealed drone brood is inserted should be strong, with plentiful nurse bees and honey; to avoid unwanted drones from polluting the mating station the selected comb ought to be protected so that the "foster" queen does not lay eggs in the cells left empty and all other drone brood present must be eliminated.

Figure 6 : **Special cage used to confine the queen to the drone comb.**

To improve success of drone production the following points are useful:

- a queenless colony is more inclined to care for drones than a queenright colony;
- removing drone brood from a colony stimulates drone oviposition;
- for the drone brood to be well cared for it is best to insert the drone comb between 2 unsealed brood combs which will attract many nurse bees;

- when sealed, the drone comb can be transferred to a any other hive in the mating station, so that the selected colony is stimulated to produce more drone brood;
- a few days before emergence of the drones it is best to feed the colonies (unless there is a very strong nectar flow).

2. MATING STATIONS

The peculiar mating characteristics of the honeybee species causes difficulties in maintaining pure races in areas where hybrids are used. A virgin queen performs a single "mating flight" in which she collects in her spermatheca* the semen from a about a dozen (8-16) different drones, enough to last for fecundation of eggs during her whole life. The matings usually take place at 30m above ground, in areas called "drone congregation area" to which drones can arrive from 10 Km or more, and queens from 1-2,5 Km. Mating stations therefore must be isolated areas in which no colonies other than the selected breeding stocks (virgin queens and drone colonies) are present. Ideal sites for mating stations are small islands or enclosed mountain valleys where the below characteristics concerning the station's location can easily be accomplished:

1) at least 5 Km distant from other apiaries;
2) situated in an area not object of transhumance honey production;

3) without natural formations favourable to setting up of wild swarms;

4) suitable for mating in climatic and geographical terms (protected from strong winds, without high populations of predators, with sufficient landmarks for orientation).

The beekeeper can further decrease risk of pollution from unknown drones by:

- providing (giving or selling at convenient price) neighbouring beekeepers (radius of 5 Km from the mating station) with selected queens;

- ensuring male saturation of the mating area by providing 15-20000 drones every 100 virgin queens;

- placing the drone colonies leeward of the nucleus boxes,

- planning the matings at the beginning or end of the season, when drone production in the area is normally low.

The mating stations can be perennial or temporary, limited to the chosen mating period. In the latter case, to avoid the presence of wild swarms, it is better to start isolation of the area before the onset of natural swarming [5].

3. INSTRUMENTAL INSEMINATION

3.1 Controlled mating

Honeybee breeders put particular emphasis on the efficiency of mating control. Natural mating of the queen can be controlled using an isolated area in which all drones - except those of the type desired

– are excluded. The isolated area does not necessarily need to be an island, but islands have been successful in obtaining pure matings. However, mating control may be less efficient than expected. Recent findings suggest that even North Sea islands do not provide absolute control [9]. Instrumental insemination (II) is an alternative to island mating and it allows the beekeeper not only to control the mating at the level of breeding stock, but also to plan specific matings: selected virgins from certain queen mothers can be mated to selected drones from certain drone mothers to give the best hereditary back ground.

Actually, the major use of II has been in research but its use in maintaining particular breeding stocks from a selected population as a first step to develop commercial breeding programs could be successfully implemented.

3.2 Rearing and maintaining queens and drones

The age of drones and virgin queens at the time of insemination is important as they must be sexually mature.

Rearing and maintaining drones

To obtain semen, drones should be at least 14-16 days or older and in healthy condition, maintained in strong colonies. If the drones are to be kept for a long time the "drone colonies" should be queenless to provide higher nursing instinct of the worker bees for the drones.

The drone colonies must be equipped with:

- a special cage (fig.7) at the entrance of the hive to prevent the drones from escaping;
- an empty super in which the drones can fly (favourable for sexual development) and from which they can be extracted (fig.7). The hive roof is opened daily to expose the super which is covered by a queen excluder so the light will stimulate the drones to fly (but without escaping !).

Figure 7 : **Hive with front cage and empty super that allow the drones to fly without escaping. The super is equipped with a hole covered with a sleeve, through which the drones can be collected.**

It is important to remember that these hives mustn't be visited during the central hours of the day as the selected mature drones could easily escape and all previous rearing efforts would vanish.

Drones can be captured and put in little cages made from queen excluders and transported as long as they are kept in a swarm box with plenty of nurse bees and honey and pollen. Once in the bee yard in proximity of the laboratory, young drones can be freed into a queenless "drone-tight" colony after being individually marked (so that they can be recognised later on, when they are mature) whereas mature drones should be immediately used.

PRE-INSEMINATION CARE OF VIRGIN QUEENS

Virgin queens should be 6 to 9 days old at the time of insemination. Queens inseminated when younger than 5 days old have showed high mortality rates and queens inseminated after 12 days of age tend to store less sperm.

Virgin queens may be emerged in nuclei or nursery colonies or incubators and, following insemination, returned to the nuclei or, in cages, temporarily to the nursery. Queens inseminated from nursery colonies should be arranged so that young larvae are next to the screen of the nursery cages [4]. It is fundamental to use queen excluders at the entrance of the nursery colonies !

3.3 Laboratory equipment and preparation of the instrument

- Insemination instrument: it consists of a heavy base supporting the hooks, the syringe holders and the queen holder (fig.8);
- Stereo microscope with a magnification of 15X to 20X. The microscope should allow adequate working space for the insemination instrument, especially for the syringe;
- Light source (preferably cold-light to avoid heating);

- Anesthetization apparatus. Carbon dioxide is used to anesthetize the queen during the insemination procedure;

- Autoclave to sterilize the saline solution and the instrument parts which have contact with the semen and with the internal tissues of the queen, to avoid contamination. In alternative a house-hold pressure cooker can be used;

- Siringe driven filter unit (0.2 μm) to sterilize the solution (in case the solution has not been previously sterilized in an autoclave);

- Laboratory centrifuge and centrifuge tubes (only in case the "mixed semen" procedure is applied).

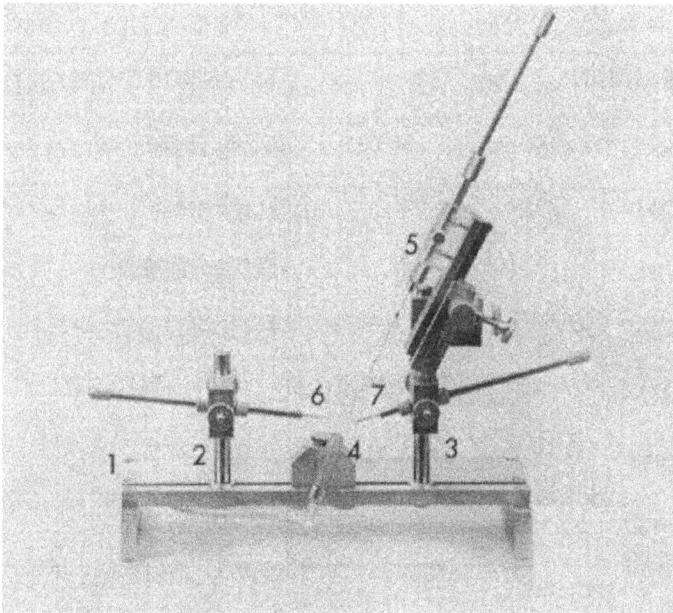

Figure 8 : **Insemination instrument by Schley [1- base; 2- left column with ball bearing-block; 3- right column additional with mechanism for insemination-syringe; 4 - queen-holder with holding-tube; 5 - insemination-syringe with tip; 6- left hook (ventral-hook); 7- right hook (dorsal-hook = hole-hook or the new sting gripping hook)]**

REAGENTS

- Diluent solution for the filling of the syringe (tab.1); in case the semen is simply collected from drones and inseminated without semen manipulation, the simplest Hyes solution is adequate or even standard saline solution (9‰ Sodium chloride) which can be bought in a chemist's or homemade.

- Sodium hypoclorite solution (15%) for the cleaning of the syringe tips;

- Chloroform for the eversion* of the drones' endophallus*;

- Formaldehyde (10% solution) or other disinfecting substance.

Tris buffer solution	Hyes solution
1.1g NaCl	0.9g NaCl
0.1g D-glucose	0.02g KCl
0.01g arginine	0.02g CaCl
0.01g lesine	0.01g NaHCO$_3$
0.48g TRIS (hydroxymethyl)-aminomethane	
0.14g TRIS idrocloride	
pH 8.8	pH 8.5

Tab.1: **The substances have to be dissolved in 100 mL of sterile distilled water. The pH value is adjusted by adding a few drops of NaOH solution.**

FILLING THE SYRINGE

The sterilized syringe cylinder is filled with sterile saline solution and the glass tip is attached to the syringe body using a 4 mm silicone sealer pushed on to the blunt end of the tip. It's important to have a very accurate control of the saline column with movement of the control knob.

COLLECTING SEMEN FROM THE DRONES

A healthy and sexual mature drone is grasped by the thorax with the thumb and fore finger of the left hand, with the drone's ventral surface upwards. The drone is easily stimulated by hand in a two step process, the partial eversion* and the full eversion. While the head and the anterior part of the thorax are squeezed with the left hand, the dorsal part of the abdomen is teased or repeatedly squeezed lightly with the thumb and forefinger of the right hand.

During partial eversion, the abdomen will contract and expose the cornua of the endophallus*. The partial eversion must be completed to expose the semen (fig.9). To obtain the full eversion, apply pressure along the sides of the anterior abdomen with the thumb and forefinger: the pressure forces the endophallus to expose the semen. The cream coloured semen lies on a layer of white mucous* (fig.9). It's very important to take special care to avoid contamination of the semen and avoid any delay in drawing the semen into the syringe because after eversion, the semen quickly spreads in a thin layer over the mucous.

The surface of the semen is then made to touch the point of the tip. When withdrawing the plunger, the semen will flow easily and fast toward the capillary tip. It is essential to separate the semen from the saline solution by a small air space to prevent dilution of the semen. Care must be taken to avoid collection of mucous as this will plug the tip and to avoid the collection of air in the column of semen. After the entire amount of semen is collected (8 μl per queen) a little air space is left before adding a drop of saline solution to avoid drying of the semen at the tip.

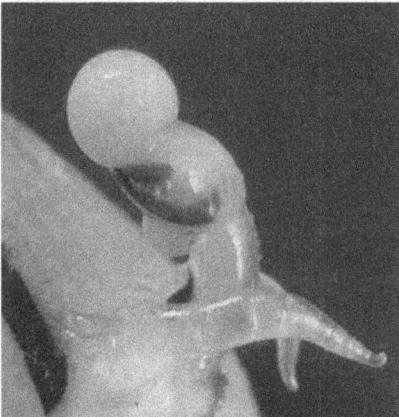

Figure 9 : **Everted endophallus. The semen is spread on the tip of the bulb on a "ball" of mucus. The two pointed appendixes are the cornua.**

3.4 The insemination procedure

QUEEN PREPARATION AND ANAESTHETIZATION

The queen should be forced to crawl into a dead-ended tube. When she reaches the end of the tube with the small hole she will backup into the holding tube held next to this. The holding tube with the queen fits on the queen holder of the instrument which is connected to the carbon dioxide line. The tube must be turned so that the back of the queen is on the right side. The queen is ready for insemination when she is motionless.

The carbon dioxide gas can be bubbled through a water bottle to determine the rate of flow. A slow steady rate of one bubble per second is recommended but the flow does not need to be too precise.

The carbon dioxide anaesthetization treatment is also a very important factor to stimulate egg laying. Two treatments are necessary to initiate egg laying, one during the procedure and a second given one day before or one day after the insemination. A second treatment of only 5 minutes is sufficient. A transparent plastic bag filled with carbon dioxide is useful to treat several queens at a time in their transport boxes.

INSEMINATION

The sting hook (or the pressure grip by Schley for the capture of the sting) and the ventral hook separate the sting apparatus from the ventral plate to expose

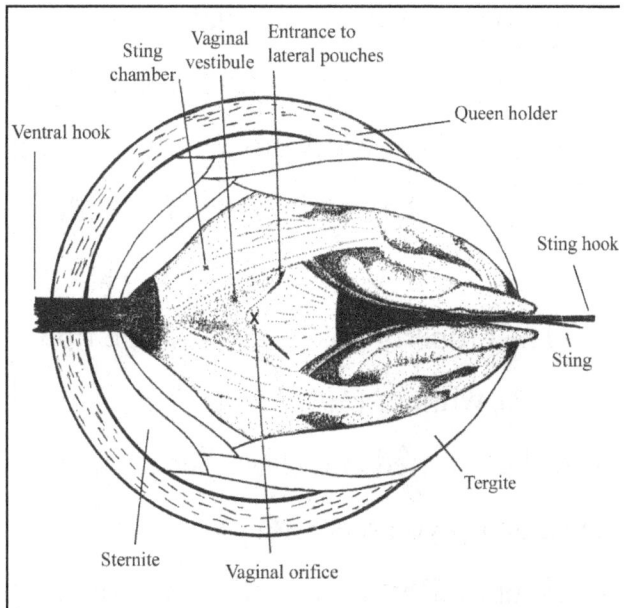

Figure 10 : **Representation of the queen's sting chamber as it appears during the insemination process (from Ruttner).**

133

the vagina. These hooks should be moved at the handles with the little-finger edge of the operator's hands resting on the table. For the beginner, a pair of forceps can also be used to open the queen allowing placement of the hooks. When the hooks are properly positioned the tissue will stretch to form a large triangle. Within this triangle is a smaller "V" of wrinkled tissue defining the vaginal orifice and location of the valve fold* which is not readily visible (fig.10). Just before insertion of the tip in the median oviduct*, lubricate the tip by dipping it in sterile saline solution The leading edge of the glass tip should be positioned above and to the right of the "V" of wrinkled tissue The tip has to be inserted about 0.5 mm (as far as the tip is wide) (fig.11-a). After the first stage of the tip insertion, to bypass the valve fold a slight zig-zag motion of the syringe tip is used, moving it slightly left, increasing the incline of the tip and inserting this another 0,5 mm into the median oviduct* (fig.11-b). The total insertion depth is about 1-1.25 mm This is essential to move the valvefold to allow for passage of the semen.

The properly positioned syringe tip will easily slip into the median oviduct without movement of the surrounding tissue. The semen can now be injected and should not leak out. If the tissue moves with the syringe tip you have not bypassed the valve fold* and the semen will back up. After insemination, the queen is marked and then returned to the nucleus.

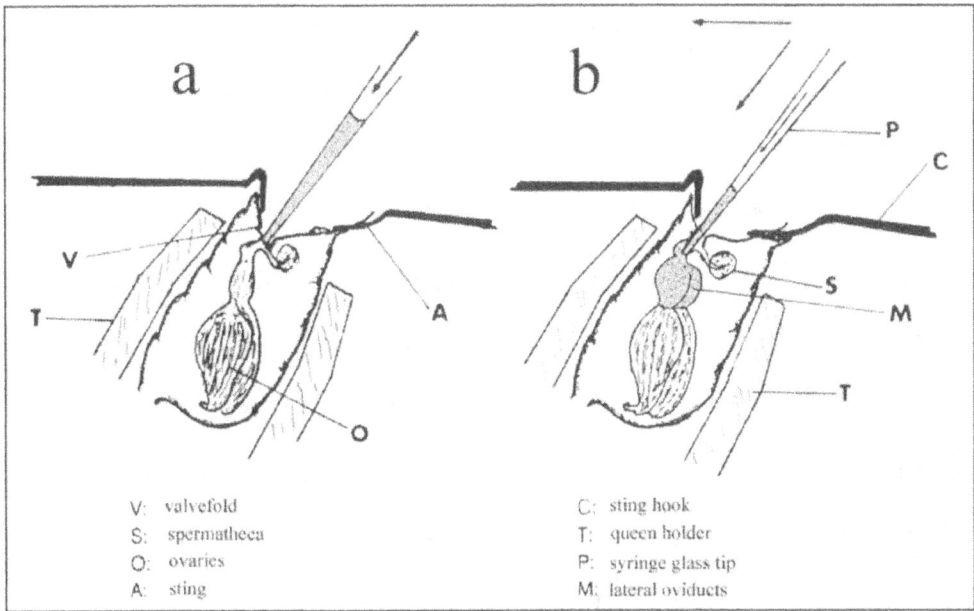

Figure 11 : **Diagram of the insemination procedure: a) the tip is inserted into the vaginal orifice for 0,5 mm; b) to bypass the valvefold the tip is moved 0,5 mm to the left, and then down a further 0,5 mm (from Harbo).**

POST-INSEMINATION CARE OF INSTRUMENTALLY INSEMINATED QUEENS

The migration of the sperm from the oviducts* to the spermatheca* is influenced by the temperature surrounding the queen during the first few hours following the insemination procedure and possibly by the activity of the worker bees and the queen herself. Having inseminated the virgin we need to put her back in her nucleus to mature into a laying queen. It's important that the method used to introduce her allows for separation for a short time but still maintaining contact or communication between the worker bees and the inseminated queen. So, queens have to be introduced carefully into their nucleus with a cage with a slow candy release to ensure the acceptance.

AMOUNT OF SEMEN AND STORAGE

One insemination with 8-10 µl or two insemination with 4 µl each are both adequate since provide enough spermatozoa reach the spermatheca of the queen: thanks to the motility of the spermatozoa and to the contraction of the longitudinal muscles in the walls of the spermatheca duct that cause a vacuum pump effect, about 5 million spermatozoa reach the spermatheca [10, 11].

Semen does not need to be used immediately after collection. It can be stored in syringe (or in tips) at room temperature (20-25°C) for as long as two days with little or no loss of viability.

3.5 Practical aspects of conservation

The insemination technique is not normally used in routine queen propagation because it requires skilled and trained specialists and special equipment; nevertheless mating control is essential to re-establish native honeybee races or populations which risk extinction in a territory where bee breeding has relied on imported stock of many races. In other countries (e.g. Italy), intensive bee breeding with the reproduction of a limited number of individuals, carried out by only a few bee-breeders and distributed throughout the whole peninsula, may have reduced genetic diversity. Selective honeybee breeding in small populations reduces genetic variance. The maintenance of genetic variation is facilitated by large population size (see Chapter 2).

Current reproduction techniques should be addressed to issues of conservation biology and biodiversity, therefore bee improvement has to be addressed by identifying desirable traits for beekeeping (e.g. disease resistance, productivity) within native bees and by population-genetic modelling designed to conserve genetic diversity. To ensure successful controlled mating, instrumental insemination is the best technique.

3.6 Special insemination techniques

MIXED SEMEN PROCEDURE

Using a mutant marker, Moritz [6, 7] demonstrated that the dilution of the drone semen and its subsequent centrifuging results in a homogeneous mixing of the spermatozoa. Among other possibilities, this technique can be used to maintain genetic heterogeneity in a breeding programme or to produce uniform -but genetically heterogeneous- insemination in a group of queens.

To get a homogeneous distribution of drone sperm in the sample, the semen has to be first blended into a tris-buffer solution (table 1) and mixed. Afterwards the sperm must be separated from the solution: a laboratory centrifuge for separation of cells and particles in density gradients is needed. This method produces uniform semen samples because spermatozoa for each insemination are taken from the same large, thoroughly mixed pool of spermatozoa (for detail see [8]).

SEMEN DILUTION

The centrifugation process of the mixed semen procedure not only places limitations on use of the technique and increases risk of contamination [3] but may also alter the composition of the seminal liquid, and thus it could cause queen loss and supersedure (Volprecht Maul, personal communications and personal experience) and a greater number of drone layer queens [1b] and supersedures in the second year of activity (personal experience).

Skowronek and colleagues [13], building on the work by Harbo [2] which describes the possibility of insemination with diluted semen – thereby avoiding the subsequent separation of the diluent by centrifugation – evaluated mixing effectiveness using drones of different races. Immediately after collection of the semen from drones of different colonies, the glass-tips are emptied into a sterile tube, to which Hyes diluent (Tab.1) is added in a 1:1 semen/diluent ratio. After careful mixing - performed with the sterilized tip of a micropipette by stirring and by repeated suction into pipette tip for 1' - the diluted semen is drawn back into the glass-tips and used to inseminate the queens with an aliquot of 8 µl. On the following day the operation has to be repeated, inseminating the same queens with a further 8 µl.. Many queen bee breeders choose the best queens only at the end of the second year, after a whole seasonal evaluation of the colonies. This means they need most of the queens to arrive at the end of the evaluation cycle. The semen mixing methodology is therefore more efficient and reliable and is beneficial from genetic and breeding points of view (implementing breeding programs that could be easily approached by queen bee breeders who need a whole productive season to carry out the evaluations). The only disadvantage of this methodology concerns the twofold insemination. This is indispensable in order to introduce into the spermatheca* the minimum dose of sperms required for the two years of activity of the queen.

References

[1a]   Cornuet J.M. (1991) Bases theoriques de l'amelioration genetique de l'abeille. These de Docteur en Sciences, 12/7/1991; Universite de Paris-Sud.

[1b]   Fischer, F. (1987) Untersuchungen zum Einfluß der Spermamischtechnik auf den Füllungsgrad der Spermatheka, in: Bericht über die Tagung der Arbeitsgemeneinschaft der Institute (1987, Mayen), Apidologie 18, 360-362.

[2]   Harbo, J. R. (1990) Artificial mixing of spermatozoa from honeybees and evidence for sperm competition, Journal of Apicultural Research 29, 151-158.

[3]   Kuhnert, T. (1988) The pros and cons of homogenising drone semen- A new technique in practical use, Proceedings of the

II° Australian and International Bee Congress, 21-26/7/1988. Published by Australian Honey Research Council, Sidney.

[4]   Laidlaw, H.H. (1979) Contemporary queen Rearing. Dadant & Sons, Hamilton, Illinois.

[5]   Lodesani, M. (2004) L'ape regina, allevamento e selezione. Avenue media, Bologna .

[6]   Moritz, R.F.A. (1983) Homogeneous mixing of honey bee semen by centrifugation. Journal of Apicultural Research 22, 249-255.

[7]   Moritz, R.F.A. (1984) The effect of different diluents on insemination success in the honeybee using mixed semen. Journal of Apicultural Research 23, 164-167.

[8]   Moritz, R.F.A. (1989) The instrumental insemination of the Queen Bee. Apimondia.

[9]   Neumann, P., Moritz R.F.A., Van Praagh J. (1999) Queen mating frequency on different types of honeybee (*Apis mellifera* L.) mating yards, Journal of Apicultural Research 38, 11-18.

[10]   Page, R. (1986) Sperm utilization in social insects, Annals Entomological Review 31, 297-320.

[11]   Koeniger G. (1986) Reproduction and mating behaviour, in: Bee genetics and breeding, edited by Rinderer T.E., Academic Press.

[12]   Ruttner F. (1988) Breeding techniques and selection for breeding of the honeybee. Published by The British Isles Bee Breeders Association.

[13]   Skowronek, W., Kruk, C., Loc, K. (1995) The insemination of queen honeybees with diluted semen, Apidologie 26, 487.

CHAPTER 6

HONEYBEE CONSERVATION: A CASE STORY FROM LÆSØ ISLAND, DENMARK.

Annette Bruun Jensen, Bo Vest Pedersen

1. CONSERVATION - WHY AT ALL?

1.1 Conservation, Rio convention

One can wonder whether the conservation of the biodiversity is really a problem impossible to circumvent, or only the dream of some scientists and retrograde naturalists. The conservation of the biodiversity can be seen as an insurance, we do not know today what will be needed tomorrow. Therefore preserving a maximum of what is still available, following the principle of precaution, is the goal. In other words high genetic diversity can be seen as an essential factor of flexibility making it possible for a system to adapt to new environmental conditions. Since the adoption of the Convention on Biological Diversity (Rio, 1992), where more than one hundred Heads of State and government met, biodiversity and genetic resources have become political buzz-words. National genetic resource programs have been established worldwide - a strong sign of political commitment and public perception.

1.2 Biological diversity and bee-keeping

Within bee-keeping biological diversity constitutes a base of work for the improvement and the safeguard of the bee of tomorrow; this base of work must be as broad as possible to offer maximum odds of success. Brother Adam and the Buckfast bee stay beyond question as an excellent example of the rational use of the biodiversity of the bee, and the future evolution of Buckfast is dependent on conservation of the different races* of bees. Many beekeepers have experienced positive achievements with well planned crossbreeding, and therefore recognized the immense value of pure races. The individual beekeeper can chose pure breeding or select cross-breeding without restrictions as long as pure subspecies/races are not disturbed and eliminated from their original geographical environment.

1.3 Resistance genes and other improvement characters

The biodiversity in the bee, it is a never-ending source of still unexploited characteristics of unknown factors that could be used for improvements. Conservation of native bee intentionally seeks to accomplish the need of beekeepers to improve their bees by breeding for diseases resistance, low defensiveness, productivity and other desirable characteristics like high winter survival.

The most illustrative example is *Varroa destructor*, which has become the most serious pest in honeybees (see the link.http://www.maf.govt.nz/biosecurity/pests-diseases/animals/varroa/maps/2000varroa-global-animap.htm). It was observed for the first time in Europe in 1971 in Bulgaria, but has rapidly spread throughout the rest of Europe. The appearance of the varroa leads researchers and beebreeders to explore the biodiversity of the bee in order to find mite resistant stocks. Thus, before the recent appearance of the varroa, nobody could imagine the mechanisms of resistance of the bee discovered today.

2. HONEYBEE DISTRIBUTION AND EVOLUTION

2.1 Natural distribution of honey bees

Millions of years ago the western honey bee *Apis mellifera* is believed to have spread westwards from Asia through Balkans and the Mediterranean region to occupy most of Europe, and southwards through the Arabian peninsula to occupy Central and Southern Africa. In this enormous territory, stretching from the Urals to the Cape of Good Hope, a large variety of habitats and climates ranging from the Continental climate of Eastern Europe with its harsh Winters, late Springs and hot, dry Summers, through Alpine, cool temperate, maritime, Mediterranean, semi-desert and tropical environments exist to which *Apis mellifera* had to adapt. These adaptations were achieved by natural selection, producing more than twenty subspecies or races* [7]. The different subspecies can be grouped into four evolutionary branches: an African branch (A), an Oriental branch (O) a North Mediterranean branch (C) and a West Europaen branch (M). Morphological, ecological characteristics and recently several different types of molecular analyses (DNA* sequences, RFLP and Microsatellites*) support this division.

The distribution of the European branches date back to the last glacial period when European honey bees presumably survived in two refugia, one on the Iberian peninsula (M-lineage) and one on the Balkan peninsula (C-lineage). After the glacial retraction 10.000 years ago the bees recolonized Europe with the M-lineage (composed of *A. m. mellifera*) occupying north and west Europe and the C-lineage occupying Central Europe (including *A. m. ligustica*, *A. m. carnica*, *A. m. cecropia*, and others). Geographical barriers, such as the Alps, maintained the differentiation of subspecies.

2.2 Manmade disturbance of the honey bee distribution

Since the 20th century natural honeybee populations in Europe have been seriously affected by human activities. Non-native subspecies of honeybees has been transported and propagated among European countries. Commercial bee breeding is and was dominated by introducing "superior" honey bees from various parts of Europe especially *A. m. ligustica* from Italy and *A. m. carnica* from the former Yugoslavia. Honey bees have a peculiar mating biology, where the queens mate in free air with numerous drones from colonies as far away as 15 km. The different subspecies or races* of honey bees can mate and interbreed, so gene flow between native honey bees, and commercially kept populations is difficult to control. As a consequence, native honey bees are considered to be extinct in many parts of Europe. Especially the black bee, *A. m. mellifera* and its various populations are endangered and must imperatively be protected, but it might also be the truth for other subspecies and races.

Brother Adam himself wrote: *"it would be an irrevocable loss if, in spite of its serious defects, the French bee indigoes came to succumb in the current of the unslung retrogression"* showing that he was also keen on the principle of precaution: one does not know what one would lose if the black bee had suddenly disappeared.

3. HISTORY OF THE LEGAL PROTECTION OF THE INDIGENOUS BEE ON LÆSØ

3.1 Honey bees in Denmark

In Danish apiculture several subspecies and hybrids*, most commonly *A. m. ligustica, A. m. carnica* and Buckfast bees, are used. Originally however *A. m. mellifera* , the black honey bee, was native and dominant in Denmark. Gradually it has been replaced by imports of the above-mentioned subspecies and hybrids favoured by the beekeepers. One of the features of the Italian bee, *A. m. ligustica* making it favourable is its superior ability to pollinate red clover. Recently it became apparent that the last survivors of the indigenous Danish *A. m. mellifera* only existed on the island of Læsø.

To keep track of the different subspecies and races* of honey bees in Denmark several of the smaller islands distributed around in Danish waters and a few peninsula are selected for controlled breeding. These locations have to be approved every year by the Danish Plant Directorate (Danish Ministry of Food, Agriculture and Fisheries) according to the Law of beekeeping No 115 article 14. In 2003 twenty-two of such locations were authorized; twelve for Buckfast breeding, six for *A. m. ligustica*, three for *A. m. carnica* and one on Læsø for *A. m. mellifera*. The latter has a further restriction only the use *A. m. mellifera* originating from the island (hereinafter 'Læsø bees') is allowed.

3.2 Læsø

Læsø is a small island of approximate 20 x 10 km (114 km^2) with a population of about 2300 citizens found in the middle of Kattegat, the sea between Sweden and the Danish peninsula Jutland. About 30 percent of the island consists of country villages, farms and small ports, while the remaining 70 percentages (84 km^2) is either uncultured consisting of moorland, marsh, tidal meadow, meadows, heath, beaches and dunes or is driven as extensive dune plantations of pine, spruce and birch. One quarter of the island resides as protected land most of which is owned by the State (figure 1). On the open area heather (*Calluna vulgaris* and *Erica tetralix*) in the northern part of the island and Sea lavender (*Limonium vulgare*) in the southern part of the island are of particular interest since they serves as the major food resources for the honey bee on the island an also provide the beekeepers with honey of higher value.

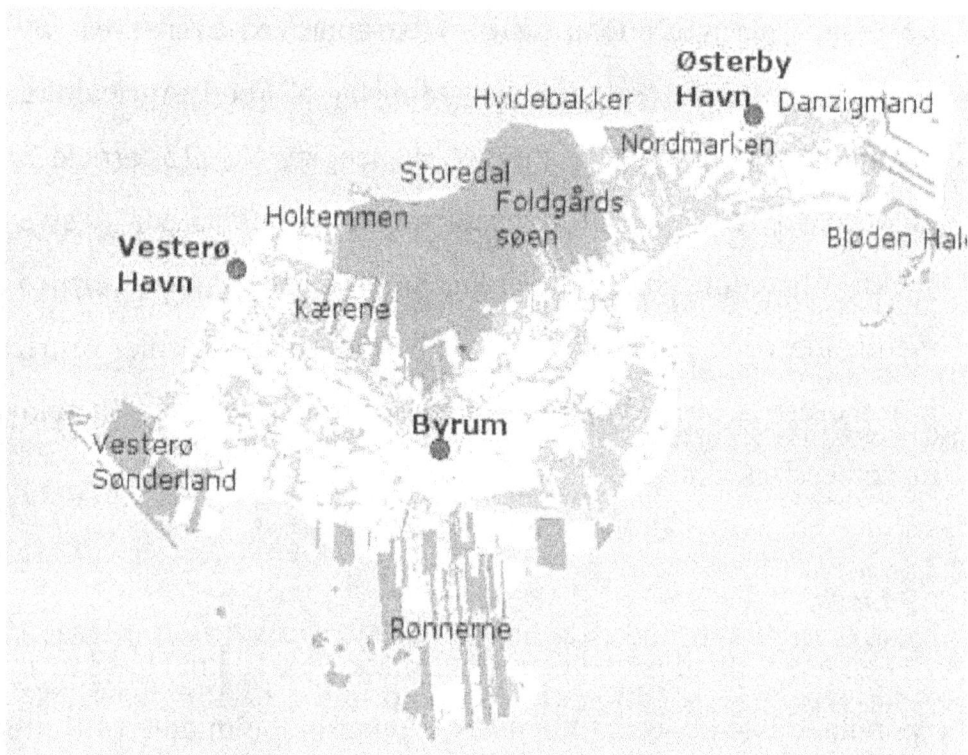

Figure 1 : **Approximate 25 percent of Læsø is nationalized, shown by grey.**

There are several factors that might explain why the *A m. mellifera* on Læsø have not yet gone extinct like in the rest of

Denmark. Over-sea migration is prevented because the distance between Læsø and the mainland is 22 km, hence foreign bees can only arrive through the act of man. The first importation of non-*A. m. mellifera* took place in the 1960'ies probably in response to the arrival of new diseases to the island such as European foulbrood caused by the bacterium *Melissococcus pluton* and Chalkbrood caused by the fungus *Ascosphaera apis*. The imported bees were not adapted to the harsh climate and the specific vegetation including the late heather bloom on Læsø and the importation was therefore limited back then.

3.3 Characterisation of the honey bee population on Læsø

The first attempt to characterize the læsø population of *A. m. mellifera* was performed in 1986 based on requests from the Læsø Beekeepers´ Association through the Danish Committee of Bee Diseases (Statens bisygdoms nævn). Morphological examinations were carried out by Danish Institute of Plant and Soil Science [8]. Cubital index* (CI) and color judgments of 116 colonies from 27 different apiaries showed that the population of Læsø bees was still rather pure; 97.4 percent of all the examined bees (c. 3000) had a CI below 1.9 and such low values are characteristic for the *A. m. mellifera* subspecies [7]. Two colonies had an CI above 2.0 and were categorized as either *A. m. ligustica* or hybrids*. In 1990 a similar survey was performed including 151 colonies from 20 different apiaries. The majority of the colonies were also this time grouped as *A. m. mellifera*, however increases in hybrid colonies were observed [9].

Mitochondrial DNA* sequences (371 bp of the cytochrom oxidase CO-I) from six of the colonies of the 1990 material and 10 additional colonies from 1992 showed that all the colonies grouped as *A. m. mellifera* or hybrids, based on CI and color, had *A. m. mellifera* derived mitochondria*. In contrast all the colonies grouped as A. m. ligusitca had *A. m. ligusitca* the derived mitochondria [1, 2]. The phylogenetic position of the Læsø bee within *A. m. mellifera* from several Europaen countries (localities in France, England, Scotland, Ireland, Norway, Sweden, Finland and Denmark) and its relationships to other subspecies has been elucidated [5, 6] by looking at differences in the base sequences of parts of the CO-I gene. All the *A. m. mellifera*, including the Læsø bee, did form a clearly distinct group in the evolutionary tree* and are more related to each other than to specimens of the subspecies *A. m. ligustica, A. m. carnica, A. m. caucasica* and *A. m. anatoliaca* (see fig. 2). Surprisingly many different maternal lineages were resolved within *A. m. mellifera* and the Læsø bee did comprise one such line alone showing its uniqueness.

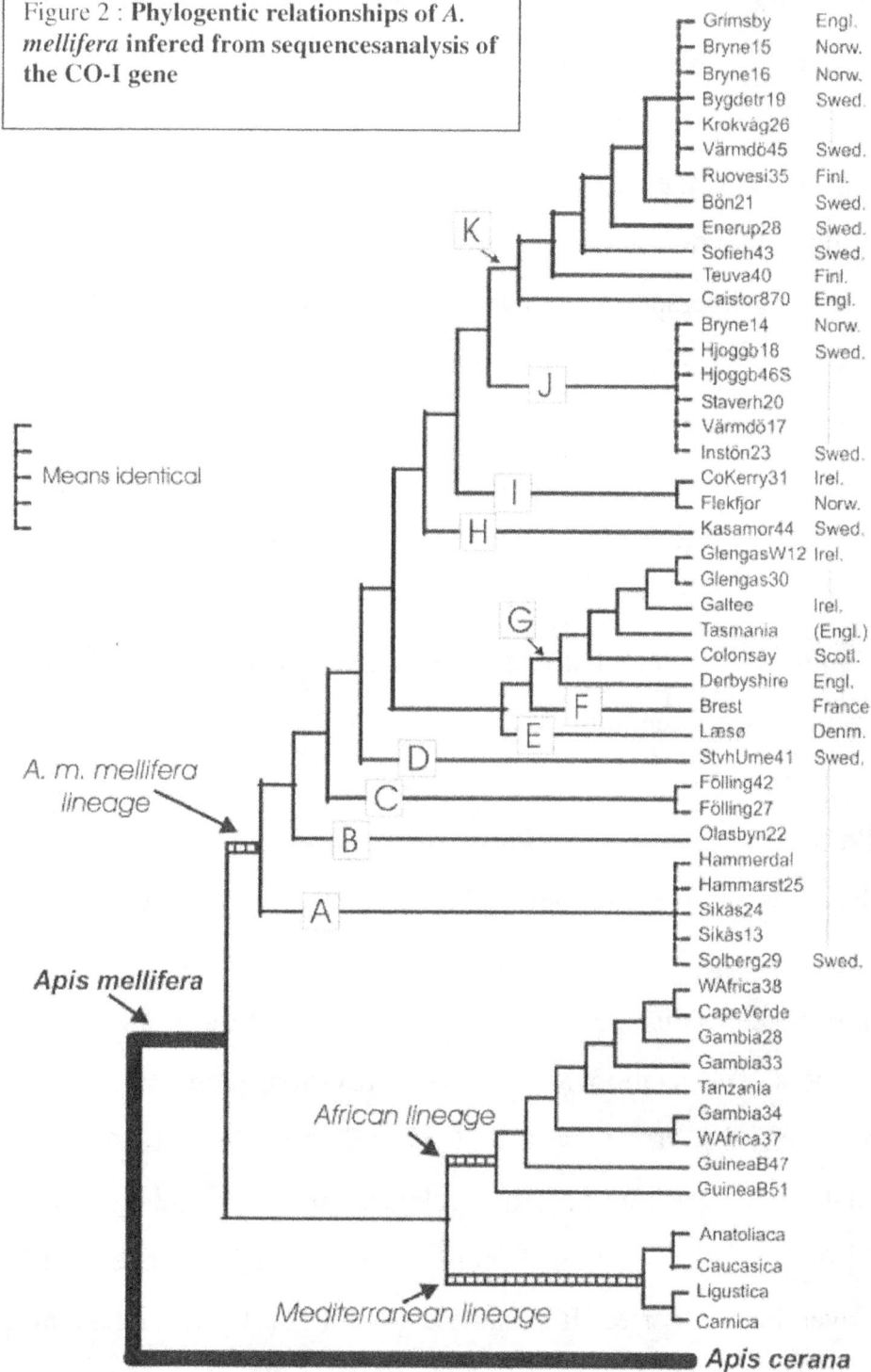

Figure 2 : **Phylogentic relationships of *A. mellifera* infered from sequencesanalysis of the CO-I gene**

Genetic markers* from the nuclear genome* (microsatellites*) and the mitochondrial* genome (haplo typing of the CO-I - COII intergenic region*) were recently used to investigate the genetic composition of the Læsø bee population and to investigate its differentiation from other *A. m. mellifera* populations in Europe. The nuclear gene diversity of the Læsø bee population were equally high with *A. m. mellifera* population from countries such as Sweden, Norway, England Scotland and Ireland, however only a single maternal haplotype* was dominant on the island. The analyses also showed that the population from Læsø was significant different from other *A. m. mellifera* populations. The distinctiveness of the Læsø population and the fact that it still pose high genetic variation making it worth a proper breeding scheme.

3.4 Protection and conservation of the Læsø bee

Based on the results from the 1986 survey a controlled breeding area for *A. m. mellifera* was established on Læsø in order to protect the Læsø bee. This measure was taken on behalf of the Danish Beekeepers' Association and its local division Læsø Beekeepers' Association, the latter was and is still the largest beekeeper association on Læsø. Not all the beekeepes on the island agreed and resulting in a long-standing dispute. Sabotage of Læsø bee apiaries have even been reported. It became apparent that the arrangement of a controlled breeding area for *A. m. mellifera* was not enough to insure the indigenous Læsø population as an increase in hybrid* bees was observed and in 1989 importation of bees to the island was banned.

In the summer 1990 a request about conservation of *A. m. mellifera* on Læsø was sent to the Minister for Agriculture on behalf of the Danish Beekeepers' Association and Læsø Beekeepers' Association.

On May the 6th 1993 the Danish government changed the "Rules of Beekeeping" in order to allow preservation of threatened honey bee subspeices like *Apis mellifera mellifera* (the Læsø bee). Article 14a of Law No 115 of 31 March 1982 on beekeeping (lov om biavl), introduced by Law No 267 of 6 May 1993, authorizes the Minister for Agriculture to enact provisions to protect certain species of bee in certain areas defined by him, and in particular provisions concerning the removal or destruction of swarms of bees regarded as undesirable for protection reasons.

In the Decision No 528 of 24 June 1993 on the keeping of bees on the island of Læsø (Bekendtgørelse om biavl på Læsøs, hereinafter 'the Decision'), the keeping on Læsø and certain neighbouring islands of nectar-gathering bees other than those of the subspecies *A. m. mellifera* (Læsø brown bee) is prohibited. The Decision also provides for the removal or destruction of those other swarms or the replacement of their queen by one of the Læsø brown bee subspecies and it prohibits the introduction onto Læsø or neighbouring islands of living domestic bees.

Full compensation from the State in respect of any loss rightfully proved to have resulted from the queen replacement or destruction of a swarm in accordance to the Decision will be provided. Finally if the removal or destruction of those other swarms or the replacement of their queen by one of the Læsø brown bee subspecies have not taken place before August the 5th 1993 it should be performed by a person approved by the Danish Plant Directorate irrespective of the owner's consent.

3.5 The judicial history of the conservation of the Læsø bee

The Danish government has with the new law including the Decision about the conservation of the Læsø bee taken a leading position concerning conservation issues in the European Union. Ideally Læsø should from 1994 and onwards host the last pure population of *A. m. mellifera* in Denmark. Unfortunately some of the beekeepers on Læsø did not approve this and, even though it was against the law, continued beekeeping with non-Læsø bees, especially yellow bees, *A. m. ligustica*.

A penal procedure was committed against these beekeepers by the Danish governments because of their violation of the national regulation. One of the beekeepers appealed to The Court of Justice of the European Union, but On December the 3rd 1998 the Court decided, that the it is legal for the Dainsh government to protect and preserve threatened subspecies like *Apis mellifera mellifera* (the Læsø bee) on an island and simultaneously ban beekeeping on the concerned island of other subspecies. The recognition of the regulation was justified under the terms of Article 36, which deals with the protection of the health and the life of animals even though it was stated that it was in conflict with the Article 30, which deals with the free interchange of goods. The appeal and the judgment of the court of can bee seen in full length on the following web page: http://www.eel.nl/cases/HvJEG/697j0067.htm.

The defendant appealed several times in national courts until January 2001 where he was convicted but during the trial no further action was taken by the authorities.

As a curiosum the citizens of Læsø have a historical reputation of being very self-opinionated. In the old land registry a huge amount of small cases reflect a rancour towards the authorities, which is similar to the above mentioned event.

4.THE PRESENT SITUATION OF THE ENTIRE BEE POPULATION ON LÆSØ

4.1 Numbers of colonies present according to the beekeepers

At present, approximately 150 colonies of Læsø bees are kept by 26 beekeepers a substantial reduction from what it was 10 years ago.

Meanwhile a large but unknown numbers of "illegal" colonies of hybrid* or non-Læsø bees are kept by 10 beekeepers. The cooperation with the beekeepers of non-Læsø bees is almost nonexistent.

4.2 Trap collections in 1998 and 1999

In order, therefore, to get an impression of the relative distribution of legal and illegal bees on the island, foraging honeybees were collected in continuant trap-collections from 10 to 13 different locations distributed on the entire island and grouped into brown bees (Læsø bees), hybrids and yellow bees (*A. m. ligustica*) based on CI and color judgments. Collection from 1998 and 1999 showed a decrease in the proportion of brown bee from 63 to 59 % and an increase in hybrid bee from 9 to 19%. The proportion of yellow bees did not vary between the two years (27-28%) [3, 4].

4.3 Foraging bees survey in 2003

In July 2003 foraging honey bees were collected directly from flowers by sucking. The island was divided into 13 different squares and approximately 100 foraging bees were collected in each square. The sampled foraging bees were assigned as brown, based on the color judgments. The proportions of brown, yellow and hybrid bees differed between the different squares but in each square some yellow bees were seen (fig. 3). 82 % of all the sampled bees were brown, which looked as a promising result from conservation point of view, but when we looked at the mitochondrial haplotype* a large fraction harbored mitochondria belonging to the Mediterranean evolutionary lineage. This was expected for the yellow and hybrids, but surprisingly more than half of the brown bees had Mediterranean derived mitochondria. The results shows that color it selves can not stand alone in assignment of a single bee to a particular group.

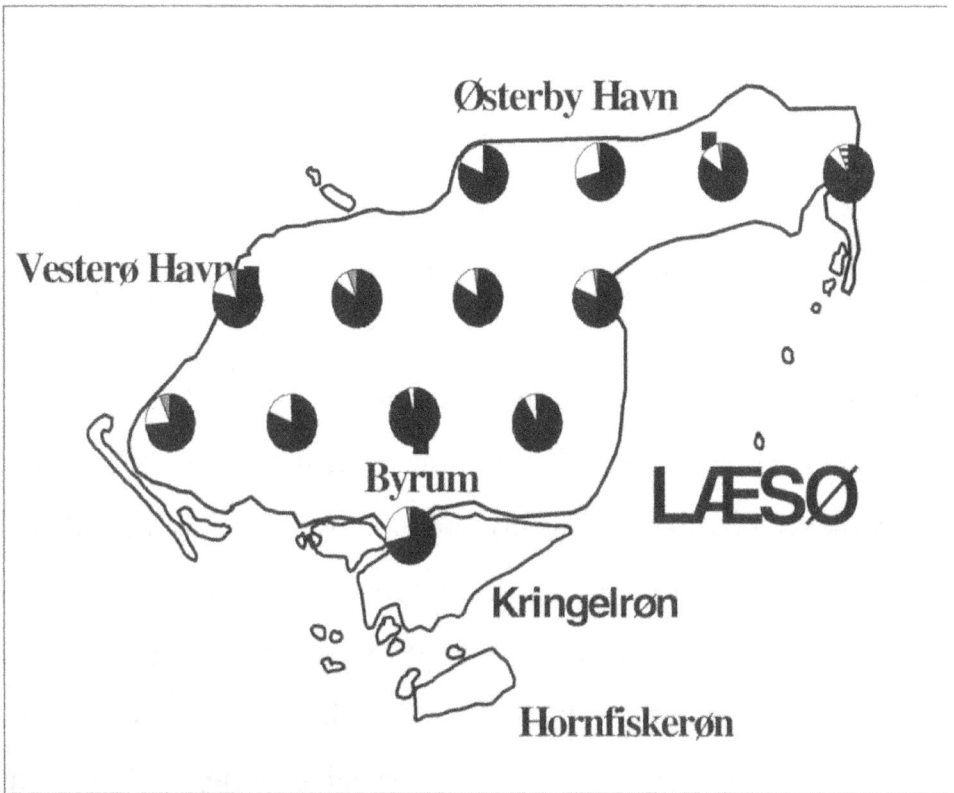

Figure 3 : **Foraging honeybees sampled on Læsø in June 2003. The pie graphs represent the fraction of black, hybrid and yellow bees based on colour alone.**

Haplotyping is not a method directly accessible for beekeepers since it requires laboratory equipment and the cubitucular index was therefore also used. A proper assignment to of the bees requires several different characters as see in table 1. When color judgments, haplotyping and CI were used for determine the purity of the bees the proportion of brown bee was only 35%, the proportion of hybrids 55 % and the proportions of yellow bees 10% (Table 1).

Even though the results cannot be directly compared to the 1998 and 1999 collections there is a clear that proportion of pure Læsø bees are decreasing on behalf on an increase in hybrid bees when we look at the entire bee population on Læsø.

4.4 The purity of the Læsø bee kept by the "Black" beekeepers

The availability of multiple molecular markers* (microsattelites) now makes possible also to investigate how much the illegal colonies have contributed to the genetic composition of Læsø bee, i.e. degree of hybridizations and introgression* of genes that have been taken place. Hybridisation can be seen as a mixing of genes between two or more distinct populations or subspecies, often populations or subspecies that after a separation are brought into contact again. This process can be mediated by man made activities or occur naturally. The genetic composition of each of the population involved in the hybridization can be described based on experimental data (fig. 4).

Subsequently the admixture of an individual can be estimated; in other words the proportion of an individual's genotypes originating from each of the involved population, or the mean admixture for the whole population can be estimated.

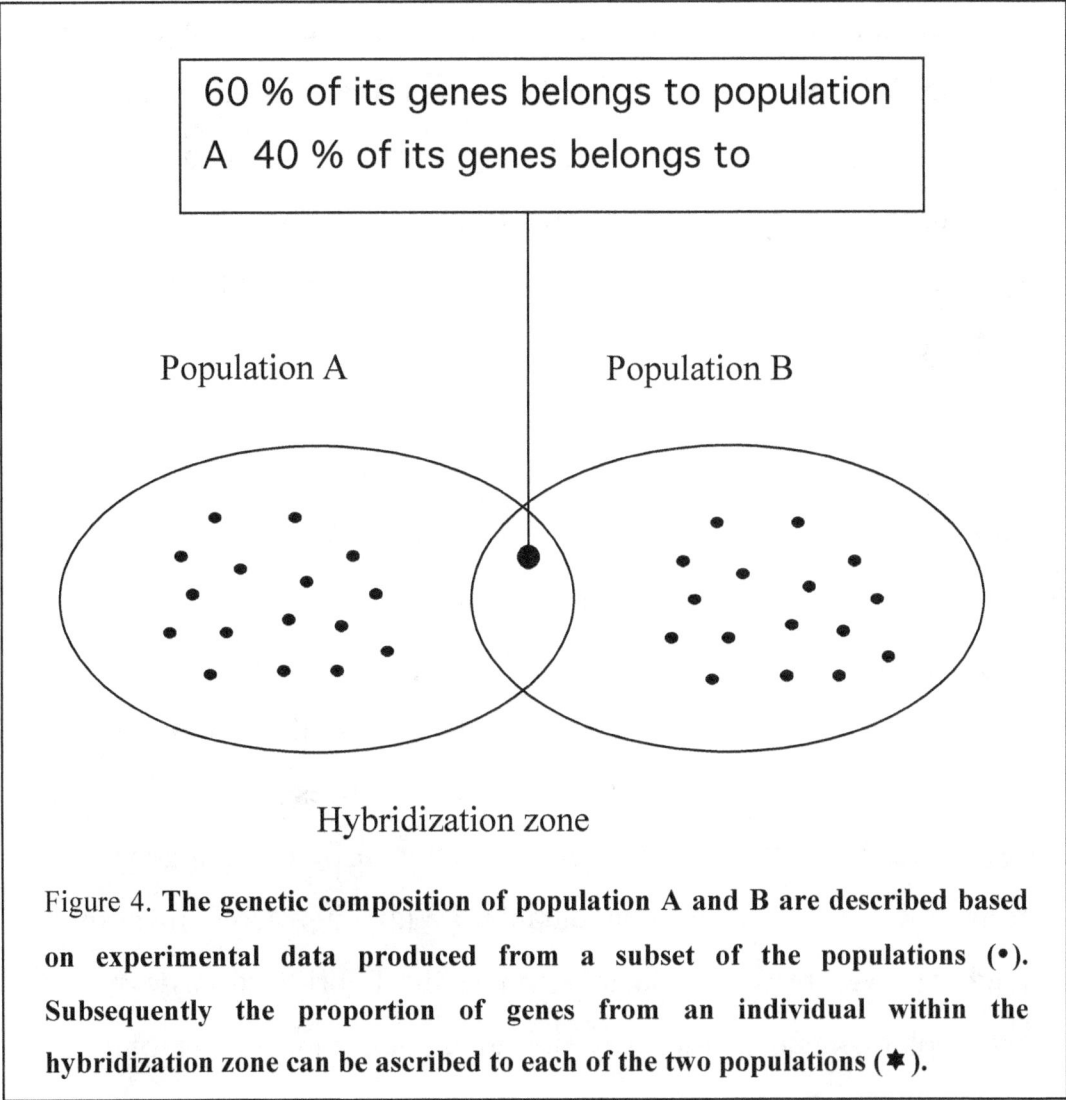

60 % of its genes belongs to population A 40 % of its genes belongs to

Population A   Population B

Hybridization zone

Figure 4. **The genetic composition of population A and B are described based on experimental data produced from a subset of the populations (•). Subsequently the proportion of genes from an individual within the hybridization zone can be ascribed to each of the two populations (✱).**

On Læsø we did not have access to the non-Læsø colonies due to the unwillingness of the beekeepers. However it is reasonable to assume that the *A. m. ligustica* bees imported to Læsø originated from the bees kept in northern Jutland, the mainland close to Læsø. Therefore we used a population of *A. m. ligustica* from northern Jutland as our reference population. We then tested the admixture between the reference population and Læsø bees collected from colonies kept by the "Black" beekeepers on Læsø. A single bee was examined per colony. The results showed that approximately one third of the bees were very pure with estimated proportion of *A. m. ligustica* genes below 2 %, whereas other bees showed moderate to highly degrees of hybridization (Figure 5).

We believe that the reason that the "black" beekeepers have managed to keep some of their colonies rather pure is a result of the awareness of the beekeepers. Several of the "Black" beekeepers in several years have excluded all colonies with just a tiny touch of hybridization for further breeding. Coloration of the worker bees and the behavior of the bees on the combs are among the characters used by the beekeepers. However due to the presences of hybrids going several generations back, this method is not sufficient and gene flow from these hybrids is unavoidable. With the above-mentioned methods, we are now able to estimate the hybridization degree of each colonies and in this way the population can rather rapidly be purified. Simultaneously however we can try to select colonies in order to preserve as much genetic diversity as possible.

5. Future

Finally after the closure of the long and principal case (in 2001) a working group has been set up by the Danish Plant Directorate to mediate a settlement between the disputing partners. The working group has proposed different models, which in general can be divided into three:

1. The Decision in maintained and removal or destruction of non-læsø bees is executed

2. Modification or suspension of the Decision in where a controlled breeding area for Læsø bees in the eastern end of the island is established. On the rest of the island colonies headed with Læsø or hybrid queens are allowed or completely open.

3. The Decision is repealed

The consequence of third model will in our opinion result in a total regression of the Læsø bees. If the Decision is repealed the beekeepers of the Læsø bees, which already have worked hard over the years to keep their bees pure, will most likely give up. If model 2 is adopted it will, with our knowledge about the mating distance and mating range require a persistent monitoring of the purity of mating.

Therefore model 1 the maintenance of the Decision is the only option if the Læsø bee shall persists and we can only recommend that it is executed as soon as possible.

In 1998 and 1999 an initiative to keep bees in South Greenland was started: The Danish Polar Bee Project. The project aimed to elucidate the beekeeping possibilities as a part-time occupation for sheep farmers. An additional aim of the project was to establish a Greenlandic population of the Læsø bees as a reference or supplement to the hopefully still existing population on Læsø. The project showed that it is possible to overwinter bees in South Greenland, however the establishment of a small population can only be expected. Therefore we cannot expect that the Grenlandic population can be used as back up for the Læsø bee

Up to now, the conservation of the indigenous subspecies or populations of honeybees has been achieved thanks to the beekeepers, professionals or amateurs. In other words, conservation could not have been implemented without the contest of the users. Therefore it is essential to interest the beekeepers in beekeeping of indigenous bees and to ensure the promotion of it if one wants to prevent its disappearance. We urge therefore to tell the beekeepers who raise indigenous subspecies or populations that they can be proud to contribute to its safeguard.

REFERENCES

[1]   Itenov K., Skjøth F., Uniformity within race and genetic constitution of a population of the black honeybee on the island of Læsø. SP-rapport 4, 1993, 35 p. (In Danish with English summary).

[2]   Itenov K., Pedersen B.V., Nucleotide sequence differences in the mitochondrial CO-I gene in the honey-bee subspecies *Apis mellifera mellifera* and *Apis m. ligustica*. Tidsskr. Planteavl 95 (1991) 101-103. (In Danish with English summary).

[3]   Münster-Swendsen, M., Fieldcaptures of honeybees on Læsø in 1998. Tidsskr. Biavl 12 (1998) 377-379. (In Danish).

[4]   Münster-Swendsen M., Bees on Læsø. Tidsskr. Biavl 6 (2000) 183-186. (In Danish).

[5]   Petersen B.V., On the phylogenetic position of the Danish strain of the black honeybee (the Læsø bee), *Apis mellifera mellifera* L. (Hymenoptera: Apidae) inferred from mitochondrial DNA sequences. Ent. Scand. 27 (1996) 241-250.

[6]   Petersen B.V., Hur många genetiske olika stammar består det bruna biet, *Apis mellifera mellifera* egentligen av? Nordbi Actuellt (2000) 21-28. (In Swedish).

[7]   Ruttner, F., Biogeography and Taxonomy of Honney bees. Springer Verlag, Berlin, 1988.

[8]   Svendsen O., Race-identification of honeybees on Læsø based on measures of the wingindexl. SP Report,1990, 14 pp. (In Danish).

[9]   Svendsen O., Bertelsen I., Meyer I.-L., Identification of honeybee-subspecies (*Apis mellifera mellifera* kontra *Apis mellifera ligustica*) on Læsø based on cubital index measurements. Tidsskr. Planteavl 96 (1992) 319-324. (In Danish).

GLOSSARY

Allele - one of the alternative forms of a gene. For example, if a gene determines the color of the thorax, one allele of that gene may produce brown and another allele produce black thorax. In a diploid cell there are usually two alleles of any one gene (one from the queen and another from the drone). Within a population there may be many different alleles of a gene; each has a unique nucleotide sequence.

Allen rule - in zoology, the principle that certain extremities of animals are relatively shorter in the cooler parts of a species' range than in the warmer parts.

Apis mellifera - the western honeybee, naturally occurring in Europe, Africa and the Middle East of Asia.

Apoidea - all bee like insects, including solitary and social bees

Base pair (bp) - a base pair refers to two bases which form a step of the DNA ladder. A DNA nucleotide is made of a molecule of sugar, a molecule of phosphoric acid, and a molecule called a base. The bases are the "letters" that spell out the genetic code. In DNA, the code letters are A, T, G, and C, which stand for the chemicals adenine, thymine, guanine, and cytosine, respectively. In base pairing, adenine always pairs with thymine, and guanine always pairs with cytosine. The length of a DNA sequence is given by the number of bp it contains.

Bergmann rule - asserts that geographic races of a species possessing smaller body size are found in the warmer parts of the range, and races of larger body size in cooler parts.

Carniolan - synonym for *Apis mellifera carnica* see there.

Caste - the females of a honeybee colony split up into two different distinct classes of individuals (=casts) the worker and the queen.

Chromosomes - the self-replicating genetic structures of cells containing the cellular DNA that bears in its nucleotide sequence the linear array of genes.

Clone – All individuals, considered collectively, produced asexually or by parthenogenesis from a single individual. Also, a copy of genetically engineered DNA sequences.

CO I or CO II - the abrivation for the genes coding for Cytochrome Oxidase I and II respectively. Both genes are placed on the mitochondrial genome and code for two subunites that unite to one of the key enzyme in the electron transport chain. Cytochrome oxidase is important for the mitochondrial ATP production and thus the energy of the cells.

Cubital index - ratio between two vein lengths of the cubital cell in the wing venation pattern.

Diploid - a cell or individual carrying two copies of a given chromosome, one from the father and one from the mother (see also *haplo/diploid population structure*). Most animal cells except the gametes have a diploid set of chromosomes. The diploid honeybee genome (queen and workers) has 32 chromosomes.

DNA (Deoxyribonucleic Acid) - the chemical that forms the basis of the genetic material in virtually all living organisms. Structurally, DNA is composed of two strands that intertwine to form a spring-like structure called the double helix. Each strand is formed by a backbone of deoxyribose sugar molecules linked by phosphate residues. Attached to each backbone are chemical structures called bases, which protrude away from the backbone towards the center of the helix, and which come in four types - Adenine, Cytosine, Guanine, and Thymine (designated A, C, G and T). Each strand of DNA has a series of Gs, As, Ts and Cs attached to its backbone. It is the sequence of these bases that forms the code which is translated by cellular machinery to create a new protein.

Drone congregation areas (DCA) - specific place where drones and queens of honeybees meet to mate in midair.

Ecotype - a population of a species that differs genetically from other populations of the same species because local conditions have selected for certain unique physiological or morphological characteristics.

Endemic - native to a given region.

Endophallus – the male reproductive organ, so called because before mating it is situated inside the drone's body.

Eversion – the process through which the endophallus is drawn out of the drone's body.

Evolutionary lineage - group of subspecies or populations that shares a common evolutionary fate through time.

Evolutionary tree - the evolutionary history or relationships of a group of organisms is often depicted as a tree. All organisms on a certain branch share the same ancestor.

Feral honeybee - honeybees living in the wild that escaped from managed colonies. This only applies to regions where honeybees have been introduced by man (America, Asia, Australia). Non-managed colonies in Europe, Africa and the Middle East should be considered as *wild* honeybees.

Gene – basic unit of hereditary material.

Gene pool - the complete genetic material of a given population or species.

Generalist - opposite of specialist. Honeybees are *generalists* because they forage on a large spectrum of floral resources in contrast to most solitary bees.

Genetic diversity - a property of a species or population, representing its complete genetic variability.

Genetic introgression - the influx of genes from one population into another by means of mating or migration of individuals.

Genome - all the genetic material in the chromosomes of a particular organism; its size is generally given as its total number of base pairs. In eukaryotes the chromosomes are enclosed in the nucleus, hence referred to as nuclear genome. In addition eukaryotic cells have mitochondria with their own genome.

Genotype - the set of two genes possessed by an individual at a given locus. More generally, the genetic profile of an individual.

Gloger rule - states that dark pigments increase in races of animals living in warm and humid habitats.

Haplo/diploid population structure - populations in which the male sex carries only one set of chromosomes (haploid) and the female two sets of chromosomes (diploid).

Haploid - organism has only a single chromosomal set (i.e. the drone has only a singe chromosomal set from its mother queen).

Haplotype - a specific combination of alleles at two or more loci in a region of a chromosome used to measure genetic variability. A mitochondrial haplotype can however refer to only one locus or to an entire mitochondrial genome. Haplotypes are a convenient way to organise genetic variation and particular haplotypes can be correlated with ecological response.

Hereditability – for a certain character, the portion of genetic variability present in a population which can be fixed by specific crosses. It is determined by the ratio between the additive genetic component and the total phenotyoic variability.

Heterozygosity - the presence of different alleles at one or more loci on homologous chromosomes.

Heterozygote - an individual carrying two different copies of the same gene.

Hybrid - the offspring resulted of matings between different species, or races subspecies.

Intergenic region - non-coding parts of a genome, which are to be found between the genes.

Interspecific competition - competition between organisms of different species.

Introgression - incorporation of genes of one species into a gene pool of another species.

Locus (plural : *loci*) – point on the chromosome where a particular gene is located.

Microsatellites - repetitive stretches of short sequences of DNA used as genetic markers to track inheritance in families. Microsatellites are the same type of genetic markers used to solve crime cases in forensic genetics.

Mitochondria - an organelle found in eukaryotes responsible for the oxidation of energy-rich substances. They are oval and have a diameter of approximately 1.5 micrometers and width of 2 to 8 micrometers. Mitochondria have their own DNA and are thought to have evolved when an early eukaryote engulfed some primitive bacteria, but instead of digesting them, harnessed them to produce energy. Offspring inherit their mothers' mitochondria, and thus mitochondrial DNA is useful in tracing maternal lineages.

Molecular (or genetic) Marker - an identifiable physical location on a chromosome (e.g., restriction enzyme cutting site, gene) whose inheritance can be monitored. Markers can be expressed regions of DNA (genes) or some segment of DNA with no known coding function but whose pattern of inheritance can be determined.

Monogynous – having only one sexually functional female in a colony.

Morphometry - the study of the structure and forms of the organism.

Mucous – a dense liquid secreted by specific glands that open into the drone's seminal duct. During eversion of the endophallus it is pushed into the bulb together with the semen, and when eversion is complete it detaches itself from the bulb and is left inside the queen as mating sign together with the chitinous plates.

Mutation - any heritable change in DNA sequence.

Nucleotide - a subunit of DNA or RNA consisting of a nitrogenous base (adenine, guanine, thymine, or cytosine in DNA; adenine, guanine, uracil, or cytosine in RNA), a phosphate molecule, and a sugar molecule (deoxyribose in DNA and ribose in RNA). Thousands of nucleotides are linked to form a DNA or RNA molecule.

Oviduct, lateral – duct leading from each ovary to the median oviduct, expandable to receive semen during mating.

Oviduct, median – the single duct leading from lateral oviducts to the vaginal chamber.

Panmictic - a population is called panmictic, when there is random mating amongst the individuals of the population, which means random choice of mating partner.

Patrilines - group of honeybee workers sired by the same drone.

Phenotype – the observable characters of an individual as determined by the interaction of its genotype and the environment.

Phylogeny - the evolutionary history of a group of organisms showing the evolutionary relationships of ancestors and descendants.

Polyandry – the mating of a female individual with several males.

Polyandrous – a female individual characterised by polyandry.

Polymerase chain reaction (PCR) - a method for amplifying a DNA base sequence using a heat stable polymerase and two 20-base primers, one complementary to the (+) strand at one end of the sequence to be amplified and the other complementary to the (-) strand at the other end. Because the newly synthesized DNA strands can subsequently serve as additional templates for the same primer sequences, successive rounds of primer annealing, strand elongation, and dissociation produce rapid and highly specific amplification of the desired sequence. PCR also can be used to detect the existence of the defined sequence in a DNA sample.

Polymorphism - difference in DNA sequence among individuals. Genetic variations occurring in more than 1% of a population would be considered useful polymorphisms for genetic linkage analysis.

Population bottleneck - a drastic reduction of population size often associated with loss of genetic diversity.

Proboscis - flexible, tubular sucking organ that extends or that is capable of being extended from the mouth of a honeybee (its tongue), used for feeding on nectar, syrup, or water, and for sampling food.

Race - group of organisms (all of the same species) that is genetically self-sustaining and isolated geographically or temporally during reproduction.

Restriction enzyme, endonuclease - a protein that recognizes specific, short nucleotide sequences and cuts DNA at those sites. Bacteria contain over 400 such enzymes that recognize and cut over 100 different DNA sequences.

Spermatheca – queen's special organ where spermatozoa are stored (5-6 million). It is linked to the median oviduct by the spermathecal duct.

Stingless bees - social bees in the tropics and neotropics that have no stinger. They can form large nests and some species are used for honey production by native tribes.

Systematics - the science of naming and classifying organisms in regard to their natural relationships (see Taxonomy).

Tandem repeat sequences - multiple copies of the same base sequence on a chromosome; used as a marker in physical mapping.

Taxonomy - the theory and practice of biological classification.

Valve fold – a muscle flap in the vagina which blocks the passage to and from the oviducts. During mating and egglaying the valvefold is folded back by the queen towards the vaginal orifice to allow the passage of semen and eggs.

INDEX